Charles H. Goodwin

Treatment of Diseases in Infancy and Childhood

Charles H. Goodwin

Treatment of Diseases in Infancy and Childhood

ISBN/EAN: 9783337371814

Printed in Europe, USA, Canada, Australia, Japan

Cover: Foto ©berggeist007 / pixelio.de

More available books at **www.hansebooks.com**

TREATMENT OF DISEASES

OF

INFANCY

AND

CHILDHOOD,

WITH OVER

Four Hundred Formulæ and Prescriptions,

As Exemplified in the Services of

Drs. A. Jacobi, J. Lewis Smith, Alonzo Clark, Austin Flint. W. A. Hammond, A. L. Loomis, W. H. Thomson, J. H. Ripley, T. Gaillard Thomas, J. R. Leaming, F. Delafield, L. A. Sayre, C. R. Agnew, L. Duncan Bulkley, Beverly Robinson, R. W. Taylor, G. H. Fox, F. N. Otis, A. A. Smith, E. C. Seguin, F. A. Burrall, E. G. Janeway, F. H. Bosworth, A. H. Smith, C. E. Billington, G. M. Lefferts, etc., etc., AND IN THE HOSPITALS OF NEW YORK CITY.

BY

CHARLES H. GOODWIN, M. D.

SECOND EDITION, REVISED

NEW YORK:

J. B. FLINT & CO.

1897.

CONTENTS.

PART I.

GENERAL DISEASES.

PART II.

DISEASES OF THE AIR-PASSAGES.

PART III.

DISEASES OF THE DIGESTIVE ORGANS.

PART IV.

DISEASES OF THE BRAIN AND NERVOUS SYSTEM.

PART V.

FEVERS.

PART VI.

SKIN DISEASES.

viii CONTENTS.

PREFACE.

In this volume the author has essayed to give a thoroughly practical, and at the same time concise, résumé of the treatment of the various diseases incident to childhood. His apology for the work must rest upon the very large number of requests from subscribers to the previous volume on Treatment of Heart and Lung Diseases.

From the numerous investigations that have recently been made in infantile diseases, and the light that has been thrown upon them, great improvement has developed in the methods of treatment. It is these recent and more advanced practical therapeutical views of the several leading medical authorities of New York City, and which in their experiences are the most approved and the most successful, that the author has endeavored to place before the general practitioner, fully and in the minutest detail and yet with the most careful discrimination.

Among medical publications the author knows of none similar in character to this, and it is earnestly hoped that the work will prove a useful one to the profession.

PART I.

GENERAL DISEASES.

ANÆMIA.

Dr. A. Jacobi believes the occurrence of anæmia in infancy and childhood, to be a circumstance of vital importance. He, moreover, directs especial attention to the fact, that although this condition is so frequently observed and so dangerous, and one which ought to be treated of by the best of practitioners and writers, yet there is no text-book in which a comprehensive study of the subject can be found.

In all cases, he advises that plenty of *out-door exercise* be taken, and if necessary this should be enforced. Crowded school-rooms, and excessive lessons, should be particularly avoided. Regarding the *diet*, irregular and fast eating is prohibited as bad habits. For infants, he directs that, in the main, solid food be withheld. Barley-water and cow's milk, in his opinion, will make far better muscle than poor mother's milk. Cow's milk, either undiluted or diluted with water only, is also prohibited. The same restrictions are insisted upon regarding condensed milk. In fact, DR. JACOBI advises that no milk whatever should be used, without the addition of some gelatinous or farinaceous decoction, such as barley-water, etc. ; and particularly in anæmia *beef-soup* should be added. At the end of the first year solid food may be given, and such articles be gradually included in the diet as physiology and experience permits.

Regarding the use of remedial agents, he finds that although *iron* has long been resorted to in the treatment of anæmia and chlorosis, yet a great many children recover in consequence of the change of diet, together with rest and an improved general nutrition, and without the use of any iron whatever. Besides, he frequently meets with cases in which the administration of iron is absolutely unavailing. In using this remedy, however, the preparations which, in his experience, are most beneficial, are the lactate, the tincture of the pomate, the pyrophosphate, the subcarbonate, and the tincture of the chloride, or muriated tincture. They should not, however, be given in the inflammatory fevers, although they may be in the septic type. Where, in addition, an absorbent is indicated, as in slow convalescence after inflammations resulting in exudation, and especially in diseases of the glands and lungs, the *syrup of the iodide* is used, thus:

℞. Syrupi ferri iodid. . . gtt.x.

Aquæ . . . . . q. s.

M. Sig. Dose, thrice daily, to a child of two years.

The iodide, he advises, meets two indications, and, moreover, the iodine is set free in the stomach and interferes with decomposition. When gastric catarrh is present and hinders the general improvement in progressive anæmia, or during slow convalescence, he finds the following combination especially beneficial:

℞. Ferri subcarbon. . . gr. iv-viii.

Bismuth. subcarbon. . gr. xii-xxiv.

Sodii bicarb. . . gr. xvi-xxx.

M. Sig. This amount daily, to a child of two years.

When the action of the heart is lowered and the blood-pressure diminished, DR. JACOBI gives the *tincture of the chloride* (being a vascular irritant) with greatest service. In cases of anæmia associated with gastric catarrh, and catarrh of the upper portion of the small intestine, he considers the *pyrophosphate* the more desirable preparation. He also uses the *hypophosphites* and the *phosphates* with good results. In chronic anæmia he finds they are all valuable. In these chronic cases, also, and especially where there is a peculiar torpid condition of the stomach, much benefit is frequently derived by him from the administration of arsenic, in minute doses, daily, well diluted with water; he gives it thus:

℞. Liq. potass. arsenit. . . . m. i-ii.
Aquæ . . . . . . q. s.

M. Sig. Dose, after meals.

Or, he often prescribes it with iron, and with or without stomachics. By this means, he finds that marked results obtain from improving the general nutrition, and this is particularly so in cases where there is also nervous trouble. He has also found *strychnia sulphate, in doses of gr.* $\frac{1}{40}$ *daily*, to a child of two years, very serviceable as an adjuvant to either arsenic or iron, its use being continued for a long period of time, combined with proper food. In some instances *phosphorus* is employed by him with much benefit, and in cases associated with caries, etc., he often obtains very favorable results. The following is the formula for the tincture of phosphorus used at BELLEVUE HOSPITAL:

℞. Phosphori . . . . gr. xxxii.
Alcohol. absol. . . . ʒ xlvi.
Tinct. vanillæ . . . . ʒ i.
Ol. aurant. cort. . . . ʒ iii.
Alcohol. absol. q. s. ad . ʒ xlviii.

* M.   Sig. Twenty minims contain gr. $\frac{1}{36}$ of phosphorus.

Concerning the benefit derived from the use of *cod liver oil*, DR. JACOBI finds that although frequently of great service, yet in his experience the contra-indications are often overlooked.   Most children, he advises, do not bear it well in the summer time, and some not at all.   In all cases, moreover, he cautions that whenever digestion is impaired, and gastric catarrh present, preliminary treatment is required before the administration of either cod liver oil or iron.

The following is DR. W. A. HAMMOND'S mixture :

℞. Ferri pyrophosphat. .   .   .   ℨ i.
    Quiniæ sulph.   .   .   .   .   ℨ i.
    Strychniæ   .   .   .   .   gr. i.
    Acid. phosphor. dil.   .   .   ℨ ii.
    Syrup. zingib.   .   .   .   .   ℥ ii.
    Aquæ ad   .   .   .   .   ℥ iv.

M. Dose : A teaspoonful or less, according to age.

DR. T. GAILLARD THOMAS prescribes the following with excellent effect for the anæmia of chlorosis,

---

* The phosphorus is digested with absolute alcohol, with exclusion of air, until dissolved ; then the flavoring ingredients are added, and finally the bulk is made up with absolute alcohol to fl. ℥ xlviii.

occurring in young girls at about the age of puberty:

℞.   Ferri vini amari   .   .   .   ℥ viiss.
     Tinct. nucis vomicæ .   .   .   ℨ iv.
     Liq. potass. arsenit.   .   .   .   ℨ ii.

M. Sig. A dessertspoonful in a glassful of water, just after each meal.

In addition to this, DR. THOMAS, regarding the indications as being to remove the cause, cure the neurosis, and repair the damages, advises general tonic treatment and the observance of good hygiene.

DR. J. LEWIS SMITH aims to improve and build up the general health of the child by the administration of the ferruginous and vegetable tonics, and by suitable hygienic measures. The following is one of DR. SMITH'S favorite tonic formulas:

℞.   Ferri et ammon. cit.   .   .   .
     Ammon. carbon.   .   .   āā gr. xxxii.
     Syrupi   .   .   .   .   .   .
     Aquæ anisi   .   .   .   .   āā ℥ ii.

M. Sig. A teaspoonful.

In cases of anæmic neuralgia occurring in children of from eight to twelve years—neuralgia dependent upon and associated with anæmia—and which he finds is due chiefly to a condition of debility, he places the child upon a most nutritious diet, consisting largely of milk and beef, and directs that it be kept away from school for a season, as the study and strict discipline exacted, together with the necessary confinement and long hours, always aggravate this malady. He also advises that plenty of out-door exercise be taken, and residence in the country secured if possible ; together with regularity in the meals, the full amount of sleep, and the avoidance of constipation.

As regards medicinal measures, he orders a moderate amount of *alcoholic stimulant*, as a teaspoonful of brandy in milk, taken at meals, with considerable satisfaction. In addition to this, his treatment consists in the administration of any of the *vegetable bitters with iron*. The following combination is a favorite with him in these cases, and from its use he has often obtained marked and usually prompt benefit ; this formula is also very extensively used at Bellevue Hospital :

R. Ferri et potass. tart. . . . 3 i.

Tinct. cinchonæ co. . . . ℥ iv.

M. Sig. A teaspoonful in a little water, four times daily.

In some instances, under this treatment, the headache, even of months standing, has ceased in less than two weeks.

DR. SMITH also derives much good from *stim-*

ulating and irritating applications along the spine.
When the pain is in the trunk, he never employs a
lengthened blister, or any blister over the spine, as
recommended by some, but usually prefers a stimulating application, as an ammoniacal mixture; or
the following is much relied upon by him for this
purpose :

    ℞. Olei terebinthinæ .   .   .   .

        Linim. saponis camphorat. .   . e. p.

M. Sig. To be applied with brisk friction, several
times daily.

Or he frequently obtains considerable temporary
relief from the use of *dry cups*, applied in three or
four places over the spine, and along its side opposite the seat of pain. But for permanent good, he
places the greatest reliance upon constitutional
remedies.

In some instances, he advises, there is also more
or less palpitation, and frequently anæmic heart mur-
murs are heard; or sometimes an occasional short
and painless cough, without expectoration, is present.
In such cases, he has found the following prescription of most valuable service :

    ℞. Ferri sulphat.   .   .   .   .   Ʒ ss.

        Acid nitric.   .   .   .   fl. Ʒ ss.

        Aquæ destill.   .   .   .   .   ℥ ss.

M. Sig. Four drops, four times daily, in sweetened water.

This cough, he finds, may also be relieved by the use of belladonna and the bromides, but at the same time advises that *iron*, *tonics*, and *careful attention to hygiene*, must constitute the chief means of treatment.

In anæmic girls at the age of puberty, and especially if chlorotic, DR. A. H. SMITH prescribes the following tonic mixture, from the use of which he has obtained the most gratifying results:

℞. Hydrarg. chlorid. corros. . . gr. i-ii.
Liq. arsenici chloridi . . . fl. ℨ i.
Tinct. ferri chloridi . . .
Acid. hydrochlor. dil. . āā fl. ℨ iv.
Syrupi . . . . . fl. ℥ iii.
Aquæ ad . . . . . fl. ℥ vi.

M. Sig. One to two teaspoonfuls in a wineglassful of water, after each meal.

This should not, however, be continued for more than two weeks at a time.

DR. ALONZO CLARK varies his treatment somewhat according to the associated condition. Besides the use of constitutional remedies, where there is also anorexia, nausea and vomiting, and constipation, he causes a free movement of the bowels, and

2

directs that they be kept in an active condition. He then endeavors to aid digestion and to improve the general health by the administration of medicinal remedies, such as *pepsin and nitro-muriatic acid*, together with good, nutritious food, and by the maintenance of good hygiene. If head symptoms are present, these, he finds, also disappear as the patient improves.

DR. BEVERLY ROBINSON prescribes the following with great satisfaction, in those cases of progressive pernicious anæmia occurring in young girls at or about the time of puberty:

℞. Ext. cascaræ sagradæ fl. . ℨ ss.

Aquæ . . . . . ℥ ss.

M. Sig. Dose, at bedtime.

This is also to be repeated in the morning if required. In conjunction with the above he then administers:

℞. Pil. ferri carbon. (Vallet) . No. x.

Sig. One three times daily, with meals.

Under this treatment benefit has followed in a very short time.

RACHITIS.

In treating these cases in infants, Dr. A. Jacobi first advises keeping the child in as uniform a temperature as possible, not allowing the room to get too hot, but maintaining an even temperature of about 70°, with a slight degree of moisture in the air by means of steam. An abundance of *fresh air* is also secured through proper ventilation ; this, he finds, is worthy of the highest consideration, and its importance cannot be over-estimated. He further directs that the child be bathed regularly and at more frequent intervals than usual, at least every morning and night. For this purpose the *cold water bath*, with perhaps a little salt in it, is used, and the child warmly clothed afterward.

Where a baby of one year is still at the breast, as is often the case, he insists that it be weaned at once, as the rachitis is probably more or less dependent upon the character of the mother's milk. In doing this, *farinaceous foods*, such as barley and oat-meal mixed with boiled cow's-milk, are gradually substituted for the breast-milk. He advises discrimination, however, in the use of barley and oat-meal, according to the condition of the bowels. If the bowels incline to be constipated, oat-meal is preferred ; but where diarrhœa is present, barley is used

because of its tendency to constipation. Besides
these, beef-tea or soup are also added to the diet.
In older children, that is to say, of two or three
years of age, he directs that plenty of *beef and eggs*
be given in addition; but where the general ill-
nutrition is very marked, with a flabby condition of
the muscles, and constipation, DR. JACOBI places the
child on a diet consisting of a few teaspoonfuls of
raw meat and one egg in the twenty-four hours,
together with plenty of oat-meal with cow's-milk.
Moreover, too much milk, he advises, may prove injurious by forming a superabundance of lactic acid
in the stomach and intestines.

Regarding direct treatment, it is questionable with
him whether any is necessary, in a great many cases.
He does not generally care to give much medicine
directly, provided he can harden and toughen the
child by cold water bathing and proper food. When
there is much cough present, opium is occasionally
required. As a rule, however, and especially where
there is more or less glandular enlargement, he gives
the *iodides*, preferring the following as one of the
best preparations:

 ℞. Syrupi ferri iodidi . . gtt. viii-x.
  Aquæ . . . . . . q. s.
 M. Sig. Dose, three times daily.

In addition the salts of sodium and calcium are

usually given in fair quantity. In many cases he also combines the above with *cod liver oil, in doses of* ℥ *i, three times daily.* The following emulsion is very much used at BELLEVUE HOSPITAL :

℞. Olei morrhuæ . . .
    Aquæ calcis . . . āāfl. ℥ viii.
    Olei cinnamomi . . . gtt. x.
M.

In cases where the child is also subject to the frequent occurrence of convulsions, or even with symptoms of meningitis present, DR. JACOBI considers it vastly more important to improve the general condition, to insure good hygiene, fresh air, and nourishing food, than to treat the convulsions, which are but manifestations of the constitutional condition. In these children, he often finds that if plenty of *oxygen*, in the shape of fresh air, is obtained, the convulsions cease of themselves without any medicinal resort. This abundance of air, he advises, must also be secured during the night, as well as the day, by means of proper and thorough ventilation.

Under this plan of treatment the results, in his experience, have been very favorable. And even in cases where there is softening of the costal ends of the ribs, with a tendency to fractures of the bones of the arms, leg, and thigh—or rather infractures—

he finds this general anti-rachitic treatment very beneficial. The fractures heal and general improvement in the condition of the child takes place. Later on, when the health is greatly improved, he remedies whatever defect may have taken place in the union of the bones, by refracturing them and permitting them to unite in a better position. Before this period, however, he often observes that all that is necessary is to bend the bones back again and retain them in their normal position.

In the treatment of all cases, moreover, caution is advised in distinguishing between rachitis due to syphilis, and those which are not; as in the latter mercurials would never be employed, while in the former they are necessary to the cure.

In treating the *sequelæ of rachitis*, found in older children, where the pigeon breast, etc., is prominent, DR. JACOBI'S main object is to expand the lungs and to change the shape of the chest. Even where the disease is of long standing, he advises that it is possible, though not very probable, that the administration of *potassium iodide* will assist somewhat in causing a resolution of the induration in the lungs, which is not infrequently present. Where, however, the lung tissues have undergone constant changes, he thinks it will prove of little value. But since similar conditions are sometimes benefited by this remedy, he recommends that it be, at least, tried in these cases; it will also act beneficially on the chronic bronchitis, which is often a troublesome feature. In administering the drug, he sometimes gives it in combination with the *iodide of iron*, for a time, with great benefit. The following method is much used at BELLEVUE HOSPITAL:

℞. Potassii iodidi . . . . ℨ iv.
    Syrupi ferri iodidi . . . ℥ i.
    Tinct. calumbæ q. s. ad. . . ℥ iv.
M. Sig. Half- to a teaspoonful.

DR. JACOBI also directs that a good, *nourishing and mixed diet* be administered, such as beef, eggs, milk, farina, oatmeal, etc.; but coffee, tea, and stimulants are strictly prohibited. *Open air exercise*, play out of doors, gymnastics, the use of dumb-bells, etc., are also required, together with cold bathing; and during the winter cod liver oil should be taken to increase nutrition.

The following is the formula for the emulsion of cod-liver oil used at ROOSEVELT HOSPITAL:

℞. Olei morrhuæ . . . . ℨ ii.
    Spts. lavandulæ co. . . .
    Spts. vini gall. . . . . āā ℨ i.
M.

DR. W. H. THOMSON finds that the best time to

give *cod-liver oil*, when regarded as a food, is about one and one-half hours after a meal, as then the stomach is acting rapidly, and the oil is passed directly into the duodenum, where it can be readily emulsified by the pancreatic and intestinal juices.

SCROFULOSIS.

Of all the remedies employed in the treatment of the strumous diathesis of young children, DR. J. LEWIS SMITH believes that the *oxide of iron* is apparently the best. This he gives in doses of gtt. i. for each year in the age of the child. In many cases of marked strumous aspect, he gives the following with great benefit :

℞. Syrupi ferri iodidi . . . gtt. iii.

Aquæ ∴ . . . . q. s.

M. Sig. Dose, thrice daily, to a child of one year.

In conjunction with this he also administers cod-liver oil. Where from the presence of an inflamed condition of the nostrils, fissures of the lips, etc., the suspicion exists that a syphilitic cachexia lies at the bottom of the case, he resorts to the application of the mercurial ointment, as follows:

℞. Pulv. zinci oxidi . . . . ℨ i.

Ung. hydrarg. nitrat. . . ℨ ii.

Oleo-paraffini (Vaseline) . . ℥ i.

M. Sig. To be applied four or five times daily.

If stomatitis is also present, he makes use of the following with much satisfaction :

℞. Potass. chlorat. . . . . ℨ i.

Tinct. ferri chlor. . . . . ℨ i.

Glycerinæ . . . . . ℥ ss.

Aquæ . . . . . . ℥ ss.

M. Sig. A teaspoonful every three hours.

In cases of scrofulous glandular enlargements, and particularly where there is also much itching present, DR. SMITH finds the following ointment very efficacious :

℞. Ung. hydrarg. nitratis . . ℨ ii.

Pulv. zinci oxidi . . . . ℨ ii.

Acid. carbol. . . . . gr. xiv.

Oleo-paraffini (Vaseline) . . ℥ iss.

M. Sig. Apply by inunction.

The carbolic acid, he advises, is usually of excellent service in allaying the itching in these cases. Or, in certain instances, he uses iodine with good effect on the adenitis, thus :

℞. Sol. iodinii co. . . . .

Glycerinæ . . . . āā ℥ i.

M. Sig. To be applied by inunction.

DR. A. JACOBI advises that in cases of strumous abscesses about the neck, after opening the abscess and evacuating its contents, he finds it an excellent plan to keep the opening patent by means of lint soaked in a two per cent. solution of carbolic acid.

DR. F. N. OTIS obtains very marked benefit, in scrofulous children, from the administration of *iodine in combination with Stuart's syrup.* The method he uses, and which he considers to be a most valuable prescription, is thus formulated :

    ℞. Iodinii   .   .   .   .   gr. xxiv.

        Potass. iodidi .   .   .   .   . ℨ i.

        Aquæ destill.   .   .   .   . ℥ ii.

    Solve et add.

        Stuart's syrup (or plain molasses) ℥ vi.

    M. Sig. A dessert- to a tablespoonful three or four times daily.

He has been accustomed to giving this to these children for many years, and in quantities that one would hardly suppose would be borne ; and which, he moreover believes, would certainly not be, if the iodine were given in any other form.

ACUTE RHEUMATISM.

———

Dr. W. H. Draper advises that the patient should be made as comfortable as possible in bed, and absolute rest secured. *Rest* he considers one of the most important therapeutic measures, and believes that no detail in this direction is too small for the personal attention of the physician. It puts the circulation in the best possible condition, and helps to avoid complications. He therefore resorts to every necessary means to obtain rest and sleep, freedom from mental and emotional changes, to alleviate the pain, and to relieve the suffering. *Position* he also considers an important factor in the treatment. The position of the affected limbs is made restful, and most comfortable for the patient. To accomplish this he finds it a good plan to put the joints up in splints, or pillows are used for supports, if necessary; this not only saves a good deal of suffering, but lessens the amount of anodynes needed. Where the pain is too severe *opium* is given, in whatever form is most agreeable. This he finds is a very valuable aid in the treatment, and advises that the benefit gained thereby should never be denied the patient.

The *diet* is another prominent feature which, in his opinion, should receive the most careful consideration. This, he advises, should be controlled in every case, both in acute and chronic disease, for by attending to the diet the recurrence of the malady may be prevented. The patient should, therefore, not be over-fed. The food is made simple, consisting of milk, which he considers best, together with animal broths, but no starches and sugars, and little bread. Gruels and farinaceous slops are believed by him to be the source of much mischief. All the water needed is allowed and, in many instances, mineral water is given, which he finds very beneficial.

Regarding medication, DR. DRAPER calls attention to the great variety of drugs which have gained a reputation as curative in this affection, due to the variation in the duration of the disease. As a rule, however, he finds that all cases will yield to the *salicylic acid* treatment, which is preferred by him. This he administers as follows:

℞.   Acid. salicylic.   .   .   .   gr. x.

Sig. Dose, every two or three hours, to a child of ten years.

Under this treatment the tenderness, pain, and discomfort usually disappear, the fever subsides, and the patient enters upon convalescence often within twenty-four hours, or in periods varying from one to

three or four days. In his experience the salicylate
treatment of acute rheumatic fever is, as a rule, uniformly successful, and no drug is shown to be so serviceable; it actually cuts short the attack, and so
long as the patient is under its influence improvement continues. He cautions, however, that the
tendency to relapse should always be borne in mind,
in using this treatment, and even after recovery has
apparently taken place, the patient is kept at rest
and the administration of the drug continued, in
diminished doses, for a period of one or two weeks;
otherwise a relapse will almost invariably occur.

Regarding complications, a great deal, he believes,
can be done to prevent them by careful attention to
treatment, by good nursing, and by a proper and
judicious diet. They should, however, be constantly
watched for, and the patient never visited without
examining for endocardial affection.

Concerning the use of *blisters*, in certain cases DR.
DRAPER applies a blister to the joint with much
benefit; but he advises that this should not be resorted to in acute attacks, where there is much swelling and fever. For, at the best, he finds that local
irritation is not of much value, but from being employed by some in these acute attacks, an erroneous
opinion is, he believes, formed of its merits. In sub-
acute forms, however, where there is a moderate de-

gree of synovial inflammation and moderate thickening about the joint, the value of blistering is found to be greatly enhanced, and excellent results are often obtained by him from their application. The following is used as an anti-rheumatic at the NEW YORK HOSPITAL:

℞. Potass. iodidi . . . . ℨ v.
Vini colchici sem. . . . ℥ i.
Tinct. cimicifugæ rac. . . ℥ ii.
Tinct. stramon. . . . . ℥ ss.
Tinct. opii camphor. . . . ℥ iss.

M.

DR. A. JACOBI places the child upon the administration of *salicylate of soda*, with very satisfactory results. The pain, fever, and other symptoms disappear, and convalescence is established often in a very short time.

DR. J. H. RIPLEY prescribes the salicylic acid treatment as follows:

℞. Acid. salicylici . . . ℨ ii.
Sodii bicarbon. . . . ℨ iss.
Aquæ ad . . . . ℥ iv.

M. Sig. A dessertspoonful every two or three hours.

DR. ALONZO CLARK sometimes uses the alkaline treatment, or in many cases he administers the *salicylate of soda, in doses of gr. viii.–x. every two or three hours* during the first day, and the same amount at longer intervals on the second day. Under this treatment he usually succeeds in controlling the disease in periods varying from one to three or four days, and in many instances within a few hours. Where much disturbance of the stomach results from its use, he gives the drug by injection with equal benefit. In all cases, however, he considers it of the highest importance to make daily examination of the heart. The cardiac disease, he finds, usually follows the articular trouble on about the fifth day; hence he advises the importance of obtaining an early control over the rheumatic affection, and thus, if possible, prevent the heart lesion.

DR. W. H. THOMSON prefers the following plan of administering the salicylate treatment, which is, moreover, a favorite prescription with him:

℞. Sol. acidi salicylici . . . ℥ ii.
    Tinct. gaultheriæ . . . ʒ i.
    Aquæ ad . . . . ℥ iv.
M.

DR. AUSTIN FLINT prescribes the salicylate treatment, using either the acid or the soda salt. In the treatment of acute rheumatism with *salicylic acid*, however, he advises that although the administration of this drug occasionally causes the disease to abort, or, at least, shortens its duration, yet it should not supersede the alkaline treatment, nor should the alkaline treatment be relinquished. This latter he carries to its full extent, as usual, since although it does not exert a marked effect on the duration of the disease, yet it diminishes the liability to pericarditis and endocarditis, which has not been shown for salicylic acid. In fact, he finds that since the use of the salicylate treatment, cardiac affections have become more common than before. This he considers to be of. the highest practical importance. Therefore, when giving this drug he also combines the *alkalies* with it, in sufficient quantity to render the urine alkaline as speedily as possible; thus:

ℹ.   Mist. acid. salicylici (gr. xlv– ℥ i)   .   ℥ ii.
     Sodii bicarbon. .   .   .   .   ℥ ii.
M.

And if this is done at the commencement of the rheumatic attack, he believes that heart trouble would in a great many instances be prevented. *Rest and quiet* are also important points in his treatment.

Moreover, while DR. FLINT places great reliance on the salicylate treatment, he has also used the *benzoate of soda* (as recommended by German authorities) experimentally in this disease. His results, however, go to show that although this agent does exert a certain amount of influence on the course of the malady, yet its effect is not to be compared to that attending the administration of salicylic acid or salicylate of soda.

At the HART'S ISLAND HOSPITAL (for children) the following anti-rheumatic mixture is extensively used ;

R. Potass. iodidi . . . . ℥ i.
    Potass. acetat. . . . ℥ iv.
    Tinct. colchici sem. . . . ℥ ii.
    Aquæ . . . . . Oii.
M. Dose : A teaspoonful.

At BELLEVUE HOSPITAL salicylic acid is very generally employed, and with excellent results. In some cases delirium is caused by its administration, particularly where a liberal exhibition of the drug has been made ; this quickly subsides, however, when the exciting cause is removed, or under the use of opiates or the bromides. In hearty, robust children of from nine to twelve years of age, the patient is put to bed and *salicylic acid gr. x. every three hours* administered; giving the following solution which, in many instances, is found to agree better with the stomach and is much used, especially in children :

℞. Acid. salicylici   .   .   .   . ℥ ss.
Sodii bicarbon. .   .   .   .   q.s.
Syrupi zingib.   .   .   .   . ℥ ii.
Aquæ ad   .   .   .   .   ℥ vi.

M. Sig. ℥ ss. contains gr. xx.

Should delirium occur after a few doses have been taken, the administration is stopped and the follow-ing given instead :

℞. Potassii bromid.   .   .   .   gr.x.
.   Aquæ .   .   .   .   .   .   q.s.

M. Sig : Dose.

This may be repeated a few times until the delir-ium ceases.

The salicylate treatment is also employed at ROOSEVELT HOSPITAL, the method being formu-lated thus :

℞. Acidi salicylici   .   .   .   gr.clx.
Sodii bicarb.   .   .   .   .
Potass. bicarb.   .   .   .   āā ℨ ii.
Syrupi limonis   .   .   .   .   ℥ i.
Aquæ ad   .   .   .   .   ℥ iv.

M. Sig. A dessertspoonful to a child of ten years.

At CHARITY HOSPITAL the following anti-rheu-matic mixture is prescribed :

℞. Sodii et potass. tart.   .   .   ℥ ss.

     Vini colchici sem.   .   .   .   ℨ ii.

     Aquæ q.s. ad   .   .   .   .   ℥ ii.

M. Dose : A teaspoonful.

SYPHILIS.

In the treatment of syphilis, mercury in some form or other is the very generally accepted remedy. The method of administration, however, is principally the variation in treatment with different practitioners.

DR. J. LEWIS SMITH, in many instances, prefers the *bichloride*, and uses the following with very satisfactory results :

℞. Hydrarg. bichlorid. .   .   .   gr.ss.

     Potass. iodidi   .   .   .   .

     Ferri et ammon. cit.   .   .   āā ℨ i.

     Syrupi   .   .   .   .   .   ℥ vi.

M. Sig. A teaspoonful three times daily, to a child three to five years old.

Or, where the general health is poor, the following prescription is found by him to be of marked benefit :

℞. Hydrarg. bichlorid. . . gr.i-ii.
Syrupi sarsaparil. co. . . ℥ ii.
Aquæ . . . . . ℥ viii.

M. Sig. A teaspoonful three times a day.

DR. F. N. OTIS gives mercury in small doses until a little impression is manifest; he then diminishes the amount for a time until this effect has disappeared, when the administration is again renewed. In giving the drug he uses *blue mass* in a great majority of. cases. In syphilis where there are scrofulous enlarge' ments, he advises that it is best to leave off the mer- curial treatment at as early a period as possible. After continuing it for some time, if the evidences of syphilis have passed entirely away and no new symp- toms present themselves, he generally omits the mer- cury and gives some milder preparation in connec- tion with potassium iodide. In these scrofulous cases DR. OTIS is in the habit of using *iodine* very freely, instead of the iodide of potassium, and finds it not only equally effective, but these constitu- tions are also braced up and benefited by it. In giving it he uses Stuart's syrup, which, in his ex- perience, is superior to any other vehicle. He has prescribed this for many years and finds it an exceed- ingly valuable preparation; the formula employed by him for this purpose is as follows:

℞. Iodinii . . . . . gr.xxiv.

    Potass. iodid. . . . . ℨ i.

    Aquæ destill. . . . ℥ ii.

Solve et add.

    Stuart's syrup (or plain molasses) ℥ vi.

M. Sig. A dessert- to a tablespoonful three or four times daily.

By this combination he is able to give iodine in very large quantity and for a long time, without any distaste and without disturbing the stomach.

The above plan of treatment DR. OTIS considers particularly valuable in cases of syphilis where there is also scrofula.

At CHARITY HOSPITAL the following mixture is very considerably used :

℞. Potass. iodid. . . . . ℨ iv.

    Syrup. sarsap. co. . . .

    Tinct. gent. co. . . . āā ℥ i.

M.

DR. R.W. TAYLOR has made the subject of infantile syphilis his especial study, and, in his experience, the following formula and method of administration are of the greatest efficacy:

℞. Hydrarg. bichlorid. . . . gr. i.

    Potass. iodidi . . . . ℨ iv.

    Syrup. aurantii . . .

    Aquæ . . . . āā ℨ ii.

M. Sig. Five drops, for a child about two months old.

This amount he then increases to gtt. xv–xx, if the disease does not yield. He advises, however, that it is important to suspend the medicine alto-gether, from time to time, as the system acquires a tolerance for it.

DR. G. H. FOX uses the *bichloride of mercury* in small doses very effectually, in cases of infantile syphilis. In tubercular forms, he often finds rapid improvement from the administration of *potassium iodide* internally, combined with the external applica-tion of *mercurial ointment.* Or, in older children and where the case is of long standing, with debility, loss of appetite, etc., he orders *mercury and potassium iodide in compound syrup of sarsaparilla*, to be taken internally, in conjunction with the local application. Should the latter cause too much irritation of the skin it is omitted, continuing the mixed treatment. Marked benefit usually follows. Frequently, how-ever, as improvement takes place the medicine causes nausea and sickness at the stomach · he then stops the mercury and administers :

R. Potassii iodidi . . . gr. ii–v.

Aquæ . . . . . q. s.

M. Sig. To be taken three times daily.

In regard to the use of mercury and other remedies in the treatment of syphilis, DR. FOX maintains that while it is a most valuable remedy, yet mercury is an overrated drug, and is not essential to the cure of the disease. He believes that it is best administered internally rather than by inunction, by vapor baths, or by hypodermic injection. In his experience, also, the amount usually given is unnecessarily large, and its local irritant effects should be avoided. As regards the duration of its use, this should vary according to the severity of the case ; no absolute rule is advised. *Iodide of potassium*, he thinks, should not be reserved entirely for the later stage of the disease, for there is no period in which either iodine or mercury is not capable of doing good. Moreover, instead of the so-called "mixed treatment," he prefers to give the two agents separately. Potassium iodide, he finds, ought not to be given continuously for any great length of time, as it acts quickly or not at all, and when unnecessarily continued is sure to do harm. Also, very large doses should not be administered without the strongest indications ; he believes they are of value in certain cases, but advises that iodism has often been mistaken for the exhibition of syphilis. *Iron* is considered by DR. FOX as deserving to be ranked with mercury and iodide of potassium, from its effects on the anæmia that invariably accompanies the early stage. *Cod liver oil* is also a remedy which he finds of great value, especially where there is a strumous condition present.

The following is found very serviceable at BELLE-VUE HOSPITAL :

℞. Hydrarg. chlor. corros. . . gr. i.
　　Potass. iodidi . . . . ℨ ii.
　　Tinct. gent. co. . . . ℥ iv.

M. .

DR. W. H. DRAPER obtains excellent results from the administration of constitutional treatment. In hereditary cases he advises the great importance of early diagnosis. The mild or the bi-chloride are often used, but he considers the *protiodide* as probably the best method of giving the drug, it being more active than the former preparation. If diarrhœa is produced, opium may be added. He also calls attention to the fact that although these patients can take mercury and iodine to an almost unlimited extent, yet the susceptibility varies considerably, which should constantly be borne in mind. When administration by the mouth is contraindicated or impracticable, he finds inunction the best method. An *ointment of mercury rubbed up with fat* is, in his experience, superior to the old form of ointment,

Sometimes tonics, quinia, etc., are also required; but, as a rule, he considers that mercury in small doses acts as a tonic, lowering the temperature and supporting nutrition. In the fever accompanying the syphilitic cachexia, and which shows itself at times with headache, etc., DR. DRAPER uses the specific treatment very effectually; the usual antipyretics, he advises, are useless, and alcohol only serves to make matters worse.

The following combination is highly esteemed at the NEW YORK HOSPITAL:

℞. Hydrarg. biniodid. . . . gr. i.

    Potass. iodidi . . . . ℨ v.

    Syr. aurant. cort. . . . ℥ ii.

    Tinct. cardam. co. . . . ℨ ii.

    Aquæ q. s. ad. . . . ℥ iv.

M.

At ROOSEVELT HOSPITAL the follcwing mixed treatment is employed:

℞. Hydrarg. chlor. corros. . . gr. ii.

    Potass. iodidi . . . . ℨ vi.

    Syrup. sarsap. co. . . .

    Aquæ . . . . . āā ℥ ii.

M.

PART II

DISEASES OF THE AIR PASSAGES.

DIPHTHERIA.

Contrary to the opinion of many specialists in throat diseases, DR. J. LEWIS SMITH considers diphtheria as primarily a constitutional disease. He believes that when the larynx is involved, the affection can only be successfully treated by solvent inhalations, or, if they fail, by tracheotomy. As a rule, he directs that the child be placed in a small room which is filled with *steam*, and the croup kettle and steam atomizer set to work. For internal medication he orders the following:

℞. Tinct. ferri chlor. . . .
　　Potass. chlorat. . . . āā ʒ ii.
　　Syrupi . . . . . ℥ iv.

M. Sig. A teaspoonful every hour, to a child of five years.

Or sometimes, in conjunction with the above, the following is also given :

℞. Quiniæ sulphat. . . . ℨ ss.

Elix. taraxaci . . . ℥ ii.

M. Sig. A teaspoonful every two to four hours.

Locally, he considers *lime* very efficient, although in many cases the progress of the disease is too rapid for its action. From experiments made in the New York Foundling Asylum, he has found that a *two per cent. mixture of liquor potassæ (or liquor sodæ) in water*, is a more efficient solvent than any other agent which is commonly used ; and with this dilution it does not cause smarting or irritation of the healthy buccal or faucial surface. It is, however, acrid and irritating in its nature, and may thus act upon those parts which are already inflamed. He therefore uses the following :

℞. Liquor potassæ . . . ℨ i.

Aquæ calcis . . . . ℥ vii.

M. Sig. To be used almost constantly, by the atomizer.

By this means the liquor potassæ, being still further diluted by the steam from the boiler, reaches the faucial and laryngeal surfaces in the proportion of one to eighty-four. In some cases, he has used inhalations of *lactic acid with acidulated liquid pepsin*, but his results have not been favorable; indeed, he does not find it so good a solvent as even the officinal lime water, and, moreover, the atomizer is apt to become corroded.

Stimulants are also given. When diphtheria complicates scarlet fever, DR. SMITH advises that more active stimulation is demanded than when the disease is idiopathic. He usually gives *brandy in doses of ℥ i. every half hour*, to a child of three years having these two infectious maladies. He has also, in many instances, employed the *antiseptic treatment* of scarlatina and diphtheria by Listerine. For this purpose, the following is a favorite with him, prescribed together with the old remedies:

 ℞. Acid. carbol. . . . m. viii.

 Liq. ferri subsulphat. . . ℥ ii–iii.

 Glycerinæ . . . . ℥ i.

M. Sig. Apply to the throat with a camel's-hair pencil, two or three times a day.

For paralysis of the uvula, which is a very frequent sequela of this disease in children, *strychnia* is his sheet-anchor. For administration, he prefers the following preparation, ℥ i. of which contains gr. $\frac{1}{80}$ of strychnia, and gives it thus:

 ℞. Elix. phosphat. ferri et strychniæ ℥ i.

 Aquæ . . . . . ℥ iv.

M. Sig. Dose, three times daily.

Under this treatment, he finds that in the majority of instances the paralysis disappears in less than a week.

DR. A. JACOBI advises that the strictest attention should be paid to even the minutest detail of treatment. He believes that in a large majority of the cases of diphtheria which terminate fatally, death occurs because of mismanagement, particularly in the worst forms of the disease; and where a case of nasal diphtheria dies, he considers it usually due to insufficient treatment or improper management. Since the adoption of cleansing and disinfecting in his plan of treatment, the results which he has obtained have greatly improved. He finds that often when there is a slight membrane, but with an enormous swelling of the glands below and behind the angles of the jaw, a sure indication of nasal diphtheria, showing that there is direct poisoning from that locality, if immediate resort is made to *nasal injections every half hour*, or every hour, marked improvement follows, and he not infrequently succeeds in reducing the glandular swellings one-half in twelve hours.

When the case is seen at the early stage with the tonsils enlarged and covered with spots of membrane, DR. JACOBI often prescribes :

℞. Potass. chlorat.   .   .   . gr. ss–i.
Tinct. ferri chlor.   .   .   . gtt. v–x.
Glycerinæ   .   .   .   .
Aquæ   .   .   .   . āā q. s. ad ℥ i.

M. Sig. Dose, every twenty minutes, to a child
of three to five years.

In addition to this, if the glands commence
to enlarge, ice is applied. He considers it important that frequent local application should
be made to the parts, and by giving the above
mixture in this way (every twenty minutes), the in-
fected tonsils are constantly washed with the potassa
and iron. At night he directs that the medicine be
given every hour; this does not interfere with rest, as
the child falls asleep almost immediately afterward.

He, moreover, lays particular stress upon the fact
that patients are left alone, often at dangerous periods
of the disease, and especially cautions against permitting the child to sleep undisturbed, a constant
tendency caused by the septic poisoning. Hence, in
these dangerous cases, he considers it highly necessary to effect a cure in thirty-six or forty-eight hours,
and advises that the child must not be allowed to
sleep during the day or night, longer than intervals
between the times of using the injections or medicines prescribed; that is to say, for a period of fif-

teen minutes or nalf an hour. Children who do not
give much trouble die, and it is these cases, he insists,
that must be thus attended to. At first he uses the
nasal injections every half hour for six or eight
hours, and then every hour. In some, however, at
the commencement he finds it necessary to inject the
nose every half hour, and to continue this for twenty-
four hours; after which once an hour may be sufficient. Careful attention to these particulars, he advises, will often result in favor of cure.

Concerning the use of stimulants, DR. JACOBI
finds that it is in these septic cases that *alcohol* is
especially indicated, but that the administration
must not be trifled with; from $\mathfrak{z}\,vi\text{–}viii.$ to $\mathfrak{z}\,x\text{–}xii.$
of brandy must be given in the twenty-four hours.
In support of this plan he mentions a case in which
diphtheria and scarlatina were combined. The child,
in spite of moderate treatment, alcoholic and other,
was sinking rapidly and almost given up; but when
stimulants in large amounts of $\mathfrak{z}\,viii\text{–}x.$ were administered, a rapid recovery followed.

As to the value of *lime water*, his opinion of this
agent is not so high as that of many writers. He
believes there is very little or no efficacy in it, since
in the amount used by the atomizer there is hardly a
trace of lime; and although the membrane is dissolved by maceration in lime water for six or eight

hours, yet he thinks this action can hardly be claimed for its momentary contact with the throat.

Regarding the internal administration of *potassium
chlorate*, he objects to the long continued and extensive use of this drug, believing that when given
in doses exceeding ʒ i. in amount during twenty-four
hours, to a child of less than three years of
age, nephritis is almost certain to result. Where,
however, catarrhal stomatitis is also present, he resorts to potassium chlorate in order that the sur-
rounding tissues may be rendered healthy. *Benzoate of sodium* he considers as of little or no service.

The following is, however, the method of administration according to the Letzerich treatment :

℞. Sodii benzoat. pur. . . ʒ i.
Aquæ destill. . . . .
Aquæ menth. pip. . . . āā ℥ i.
Syrup. aurant. cort. . . . ʒ ii.

M. Sig. A dessertspoonful every hour, to a
child under one year.

His opinion of the utility of *pilocarpine* is also unfavorable, and in those cases in which the deposit is
deeply imbedded in the lower structures, he believes
it does positive harm, possibly hastening a fatal issue.
If the drug be used at all he prefers the fluid extract.

as the muriate of pilocarpine is decomposed in the stomach.

In regard to the *removal of the membrane*, so far as the tonsils are concerned, DR. JACOBI objects to this interference; in other situations, however, he removes the false membrane when possible to do so. But he cautions against resorting to this procedure injudiciously, and prefers to continue the injections and washings for a long time, rather than to attempt the removal of membrane which is not perfectly loose. Finally, in all cases, he urges that careful attention to *nourishment* is of the greatest importance. He directs that the child be abundantly fed with the most nutritious diet, given largely every hour or two, in a liquid form; and that life be sustained by stimulants and nourishment during the night in the same manner. By this means he believes some lives may be saved that would otherwise be lost.

In the treatment of the *sequelæ of diphtheria*, DR. JACOBI advises that the majority of cases of paralysis run a benign course, and get well in eight or ten weeks. His treatment is in the main tonic and stimulant, feeding the child on beef, milk, and eggs, with alcohol in small quantities. As a tonic *iron* is preferred, varying the preparation according to the case. If diarrhœa is present, he uses the sulphate of iron most beneficially. For a nervine he considers strych-

4

nia as most efficient and powerful in its effects, and
gives it to the amount of gr. $\frac{1}{25}$ daily, to a child of
four years, beginning with *strychniæ sulph. gr. $\frac{1}{80}$ three
times a day*. When the paralysis is extending, and
fear is entertained that the respiratory muscles may
become involved, he resorts to the *galvanic and faradic currents*, principally the latter, and administers
hypodermic injections of strychniæ in doses of gr. $\frac{1}{80}$
once or twice, or several times, daily, with great benefit. By these measures, together with the observance of
good hygiene, he usually effects a cure in most cases.

Where the disease is complicated by nephritis, he
orders the child put to bed and administers either
the *hot bath* or hot pack twice daily to produce diaphoresis. Or, where the heart's action is good, he
sometimes induces a free perspiration by means of
hypodermic injections of *pilocarpine*, given two or
three times in the twenty-four hours ; and in addition
to this *alcohol* is employed to counteract the depression on the heart caused by the pilocarpine. In this
way he occasionally gives as many as eight injections
in twenty-four hours, always seeing that the child is
well stimulated ; this extreme resort is, however, not
advised in all instances, each case requiring its own
special treatment. Sometimes DR. JACOBI uses
gallic and tannic acid, in doses of gr. v-xv. daily, in
conjunction with the pilocarpine, and often with very

favorable action. The bowels are also kept gently open by means of the cautious administration of the salines. Toward the later stages he finds iron very serviceable when there is much albumen in the urine.

DR. C. E. BILLINGTON considers diphtheria as primarily a local affection, and that its specific constitutional phenomena are secondary results by absorption. He therefore believes *local antiseptic treatment* to be the most efficient means of successfully combating this disease. In employing this treatment, however, he at the same time cautions that the danger to be most avoided is irritation ; this, he advises, may render the method unsuccessful by the evil effects which it causes. To remove the foul secretions from the throat, the formulæ which he habitually uses are as follows :

℞. Potass. chlorat. . . . . Эii.
    Glycerinæ . . . . . ℥ss.
    Aq. calcis . . . . . ℥iiss.

M. Sig. A teaspoonful every half-hour, to a child of two to five years.

This he usually administers in alternation with the following :

℞. Tinct. ferri chlor. . . . Ʒi.
    Glycerinæ . . . . .
    Aquæ . · . . . . āā ℥i.

M. Sig. A teaspoonful every half-hour.

Both of these mixtures, he advises, are pleasant to the taste, easily taken, and cause not the slightest irritation to the parts. In some cases he omits the latter with advantage. In addition to these measures, he also sprays the throat very frequently, when practicable, and for some minutes by means of the hand atomizer, using the following solution:

℞. Acid. carbol. . . . . m. x.
Aq. calcis . . . . ℥ iv.

M. Sig. To be used every half-hour, by the spray.

Regarding the use of the *spray of carbolic acid and lime water*, DR. BILLINGTON considers it to be a remedy of great value, not only in its solvent action on the false membrane, but as a most agreeable and efficient aid in subduing inflammation. He also finds it one of the most reliable means for efficiently applying disinfectant treatment to the larnyx ; and in his experience it not only sometimes cures, but tends to prevent laryngeal diphtheria. In these laryngeal cases he uses it as nearly constantly as practicable. The careful attention to all these details cannot, he advises, be over-estimated. Finally, except in bad cases, when he considers this first in importance, he reaches the affected surfaces at suitable intervals by means of a syringe, using a weak, tepid *solution of common salt*. This he throws through the open mouth into the throat, or, where the nares appear to be invaded, as shown by a discharge from the nose or by obstructed breathing, through the nostrils into the nasal passages and the pharynx; and continues the process until the foul secretions are thoroughly washed away and the fetor is corrected. The use of throat and nasal syringing is regarded by him as of great value, and he believes, moreover, that not a few cases terminate fatally where these measures are not employed.

Regarding internal medication, in the treatment of young children he finds it very important that the employment of any unpleasant medicines, such as *quinine*, *cubebs*, etc., should be particularly avoided. As to the administration of *alcohol*, he considers it useful in the later stages of protracted cases, and during convalescence. The prevalent custom, however, of a free and early resort to stimulants he believes to be injurious, and it is his experience that they often add to the poison of the disease, and help to overwhelm rather than to sustain; therefore in a great many instances he has treated absolutely without their use. In all cases, DR. BILLINGTON insists upon proper *nutrition* as being of the highest importance. Milk is preferred by him, given freely and often, and even using force if necessary; when indicated, also, he advises that rectal alimentation should never be omitted.

By this means of treatment, carefully and judiciously carried out, and with the strictest regard for every minute detail, the results obtained by him have, as a rule, been most gratifying.

DR. A. A. SMITH finds that *potassium chlorate in doses of gr. i every half-hour*, will produce the same results as larger doses, and without the danger of evil effects ensuing from accumulation of the drug in the system, which sometimas happens when administered in the ordinary way. Indeed, he believes this method will give more beneficial effect on the throat inflammation. On this plan he usually prescribes:

℞. Potass. chlorat. . . . gr. xvi.

Aquæ . . . . . ℨ ii.

M. Sig. A teaspoonful every half-hour.

DR. FRANCIS DELAFIELD considers the internal administration of *alcohol* and the use of *iron and potassium chlorate* as probably the best means of treatment. He gives alcohol, however, not merely for sustaining the strength of the patient, but for its constitutional effects ; hence in large doses and in concentrated form, using *brandy or whiskey in doses of ℨ i. every hour or two*, and to the saturation of the system. By this means he also finds that the drug acts as a cardiac stimulant after the disease begins to improve. He advises, moreover, and especially in severe cases, that the use of alcohol must be commenced early. The iron and potassium he gives as follows:

℞. Tinct. ferri chlor. . . gtt. v-x.

Sig. Dose, every two hours.

This is alternated with:

℞. Potass. chlorat. . . . gr. x-xv.

Sig. Dose, every two hours.

These are administered so that the child gets one every hour. The use of *soda benzoate in large doses of gr. x-xxx. every three hours*, he also considers an excellent plan of treatment in many cases.

The local measures adopted by him consist in the application, in some instances, of *carbolic or salicylic acid* by means of the brush or spray; or, in others, of the *tincture of the chloride of iron*. This, he advises, must be made every two or three hours, to obtain good effects. *Sprays, steam inhalations*, and *vapor of lime*, are also used at times. Where, however, the larynx is involved and suffocation is imminent, *tracheotomy* is indicated, although he finds it even less satisfactory than in croup.

Dr. BEVERLY ROBINSON frequently uses lactic acid in the form of a spray with much benefit. The solution he employs is as follows:

℞. Acid. lactic. . . . . m. xx.

Aquæ . . . . . ℥ i.

M.

DR. J. H. RIPLEY considers the vapor of *lime water* as of little or no service, since the patient gets so little of it. He has used it in the form of the spray, and poured it by teaspoonfuls into the trachea, but the results have been unfortunate.

DR. ALONZO CLARK uses the spray of lime water as frequently and as constantly as indicated, placing much reliance on its efficacy when thrown upon the forming membrane.

DR. L. ELSBERG, has found *bromine* to be a much more efficient solvent for false membrane, than either lime water, lactic acid, liquor potassæ, or liquor sodæ ; and if properly used, he finds that it does not cause any irritation. For this purpose he prescribes the following :

℞. Brominii . . . . gr. i.
Potass. iodidi . . . ʒ i.
Aquæ . . . . . ℥ i.

M. Sig. For inhalation.

This, he advises, can be poured into a cone, and the child be permitted to inhale the fumes. In his experience this solution has better solvent effect on the membrane outside, and apparently inside also, than either of the above agents.

DR. J. R. LEAMING has used *calomel* very effectually in many instances. This drug, however, he administers in large doses, and has frequently obtained the best results in a very few hours.

DR. E. C. WENDT has prescribed the muriate of pilocarpine with success in many cases, where other remedies had been tried without the least benefit. This he gives, as advocated by Guttman, by the following method, which has been found most advantageous:

R. Pilocarpin. muriat. . gr. ⅓–⅔.

Pepsin. . . . . gr. i–i¼.

Acid. hydrochlor. . . . gtt. ii.

Aquæ destill. . . . ℥ iiss.

M. Sig. A teaspoonful every hour, to a child of eight or ten years.

In conjunction with the above, each dose is followed by *teaspoonful doses of old sherry wine.* By this treatment, kept up continuously throughout the night, he has, in many severe cases, observed marked improvement in a very short time.

* The ALCOHOL, or so called "BROOKLYN," TREATMENT of diphtheria, which has yielded such excellent results, consists in the administration of alcohol, not, however, as in a variety of other diseases, to relieve or prevent great prostration, but for its specific action as an antidote to the diphtheritic

---

* The Brooklyn Treatment of Diphtheria may be considered as of sufficient importance to entitle it to an insertion here.

poison. In these cases its stimulant or other ordinary effects are not induced by it, and intoxication
has never been observed, while enormous doses have
been administered to young children. Not only,
moreover, as an antidote, but also as a prophylactic
of diphtheria is the value of alcohol demonstrated,
when administered early and in sufficient doses. For
this latter effect the following prescription is very
generally employed :

    ℞. Cinchon. sulph.   .   .   .   ʒ i.
        Acid. sulph. aromat.   .   .   . q. s.
        Spts. vini gall.   .   .   .   .   ℥ vii.
        Glycerin. opt.   .   .   .   . ℥ i.

    M. Sig. From a small teaspoonful to a table-
spoonful every two hours, according to age.

    And even to prevent or cut short any of the se-
quelæ, it is claimed for this plan of treatment that
there is no remedy so reliable as alcohol. In the
majority of instances *quinine* is also administered,
although in many cases alcohol alone is employed,
and the effect seems to be almost as decided and
quick without, as with, quinine. Where the medicine
is refused on account of the taste of the drug, the
alcohol is often given separately, and the quinine
administered by inunction, as follows :

    ℞. Quiniæ sulph.   .   .   ˙   .
        Chloroformi

Bals. peruv. . . . . āā ℥ ss.

Adipis . . . . . ℥ vi.

M.

For the alcohol, the following formula is usually employed:

℞. Spts. vini gall. . . . ℥ iiss.

Glycerinæ opt. . . . .

Syrupi simp. . . . . āā ℥ ss.

Aquæ menth. pip. q. s. ad . . ℥ iv.

M. Sig. A teaspoonful every hour, to a child under two years. From two to three years, ℥ iss., from three to five years, ℥ ii., and from five to eight years, ℥ iiss.– ℥ ss. every hour.

Or, in many instances the following combination may be used very satisfactorily:

℞. Quiniæ sulph. . . . gr. xvi.

Acid. sulph. aromat. . . q. s.

Spts. vini gall. . . . . ℥ iss.

Syrup. simp. ad . . . ℥ ii.

M. Sig. A teaspoonful every two hours.

In addition to the above, whiskey may be given in doses of ℥ i–iii. as indicated. *Champagne* is also used in many severe cases.

The local measures resorted to consist in the application of a light astringent, either in the form of a gargle or spray, or as a powder to be blown directly on the diseased parts ; *chlorate of potassium, powdered*

alum, and *tannic acid*, either combined or otherwise,
with the addition of a comparatively large quantity
of *carbolic acid*, being used for this purpose.

Besides these means of treatment, great importance is always attached to *hygiene ;* cleanliness and
fresh air commanding primary attention. In regard to
nourishment, during the first twenty-four hours of
an attack little or none at all is given, as a rule, but
in all cases the first food administered consists of
milk and lime water, with the addition of a little
salt. During convalescence iron, particularly *dia-
lysed iron*, is employed, and, in combination with some
form of alcohol, either prevents or relieves the conditions of anæmia and loss of nerve power which
are so apt to follow diphtheria.

CORYZA.

In acute cases, where the attack is of a mild character, Dr. J. LEWIS SMITH advises that very little
treatment is required. He directs that the bowels be
kept open, the feet soaked in mustard water, and the
body warmly clothed. *Inunction of the nostrils* is
also resorted to, and often with much relief. Where

there are evidences of the extension of the disease toward the bronchial tubes, he gives an emetic of syrup of ipecac, followed by:

R. Syrupi ipecac.   .   .   .   . ℥ ii.
Spts. æther. nitr.   .   .   .   . ℥ i.
Syrupi simp.   .   .   .   . ℥ ii.

M. Sig. A teaspoonful every three hours, to a child of six months.

He further employs *injections* very beneficially. A three to five per cent. solution of common salt in warm water is used, injecting it into the nostrils with a small syringe. This aids materially in removing the muco-pus which obstructs the respiration, and in establishing a healthful state of the inflamed surface. Or, sometimes a mixture of carbolic acid, glycerine and water, with a few grains of potassium chlorate added, is used instead of salt.

Chronic cases of coryza, or "snuffles," are usually attributed to the strumous cachexia, and often associated with glandular enlargements, etc. For local measures, in these patients, DR. SMITH treats by injecting into the nostrils with a glass, or hard rubber, or rubber ball, syringe, a solution of equal parts of lime water and warm water, at a temperature of about 100°, the child being placed on the back with a towel laid over the eyes. Internally he gives the following:

℞. Olei morrhuæ . . . ℥ ii.

Syrupi ferri iodidi . . . ℨ i.

M. Sig. A teaspoonful three or four times daily, to a child of one year.

If constipation is present, which is frequently the case, he finds this doubly serviceable, since the cod liver oil will also act as a laxative besides its effect on nutrition. In addition, and especially in chronic cases associated with syphilis, he considers the following an excellent application to parts which can be reached by a camel's-hair pencil, or with a nasal sponge :

℞. Ung. hydrarg. nitrat. . . . ℨ ii.

Ung. zinci oxidi. . . . ℥ ii.

M. Sig. Apply three or four times daily, as far within the nostrils as possible.

Where the child is anæmic, he further advises that a teaspoonful of the juice expressed from rare beef-steak be given every two hours.

Dr. A. A. SMITH often uses *aconite* very beneficially. In many cases seen at the commencement, with more or less fever, and where the skin is hot and dry, the pulse full and bounding, and the mucous membrane of the throat and nostrils dry, he gives the following with decided benefit :

℞. Tinct. aconit. rad. . . m. $\frac{1}{3}$–$\frac{1}{2}$.

Aquæ . . . . . q. s.

M. Sig. Dose, every fifteen minutes.

When perspiration appears, which usually occurs in a short time afterward, he then administers the drug at longer intervals ; every half-hour or longer, according to indications.

In older children, the following is recommended very highly by DR. J. R. LEAMING, as a means of aborting a cold if taken early:

℞. Ammon. chlorid. . . .
    Potass. nitrat. . . . .
    Senegæ . . . . . āā ℥ ss.
    Glycyrrhizæ . . . . ℥ i.
    Aquæ . . . . . . Oi.
M. Sig. ℨ ii to ℥ ss. every half-hour.

TRUE CROUP.
(ACUTE CROUPOUS LARYNGITIS.)

In the treatment of this disease, Dr. J. H. RIPLEY considers the use of medicinal measures exceedingly unsatisfactory. The only rational plan, in his experience, is a symptomatic one. Stimulants and extra feeding are, however, required in severe cases. If seen in the early stage before serious dyspnœa is present, he at once applies a large *warm poultice* over the larynx and extending nearly down to the root of the neck ; it is also made to fit as closely as possible without obstructing the circulation or respiration, and changed as soon as it becomes cool. This, he finds, not only affords comfort but in many cases gives, at least, temporary relief. The temperature of the room is kept at 70° or 80°, and the air constantly moistened with steam. In addition, if possible, he also secures direct application of *steam* by means of the croup-kettle, using it five minutes at a time at intervals of fifteen minutes, and finds that it often renders most valuable service. While the child is sleeping, a gentle stream, not too warm, may also be kept up continually. As regards *lime water*, he does not find it any more serviceable than steam, whether produced from slaking lime, or thrown by means of the spray. The spray is, however, used in many cases, and the atmosphere saturated with lime steam by means of kettles filled with a mixture of quicklime and water, and kept boiling.

Concerning *emetics*, DR. RIPLEY prefers either the sulphate of copper or the yellow sulphate of mercury, as most prompt and efficacious. In using them he tries the drug once or twice, but no more, as he decidedly objects to the repeated exhibition of these depressing agents, thus exhausting the patient. In addition to the above means of treatment, as a rule, he prescribes the following :

℞. Potass. chlorat.   .   .   .   . gr. v.

Tinct. ferri muriat.   .   .   .   gtt. v.

Syrupi   .   .   .   .   .

Aquæ   .   .   .   .   . āā q. s.

M. Sig. Dose, every two hours, to a child of one to five years.

This is, in many instances, continued throughout the whole course of the disease. Should diarrhœa result from its use, as sometimes happens, the administration is stopped and, if required, an emetic is given. *Stimulants*, brandy and milk, are also used as indicated. Where there exists a previous history of malaria, and with high temperature, quinine is ordered ; in some instances giving ;

℞. Quiniæ sulph.   .   .   .   gr. xx.

Sig. This amount during the twenty-four hours, to a child of two or three years.

Or, in others, gr. v. is administered, and the dose repeated in one or two hours. In older children gr. x. may be given at night, followed by gr. v. in the morning, and this amount increased as demanded.

Sometimes, instead of the potassium chlorate and iron solution, he finds bromine very serviceable, thus :

5

℞. Brominii   .   .   .   .   .   ℥ss.

Aq. cinnamomi   .   .   .   . ℥iv.

M. Sig. A teaspoonful every two hours, taken in milk.

This may also alternate with stimulants in the form of milk punch. He advises, however, that the bromine often causes marked gastric irritation, and cannot be continued with regularity. If the case is seen early and patches are discovered on the tonsils, he sometimes cauterizes them with a *solution of silver nitrate*, and prescribes bromine internally, as above.

When, however, all these measures are unavailing, and medicinal treatment has failed either to relieve or to check the progress of the disease, and with exhaustion and cyanosis approaching, DR. RIPLEY immediately resorts to *tracheotomy*. In such a case the after treatment, he advises, requires the utmost care and attention. The child should be seen every eight hours, for the first few days following the operation. ˙ Usually during the first twenty-four hours nothing unfavorable occurs. If the breathing is perfectly free, he fastens a large moist sponge over the mouth of the tube, which accomplishes the two-fold purpose of warming and of moistening the air, and can easily be kept clean, while it also allows of a ready manipulation of the tubes. Should the breathing be at all harsh, he directs that the *inhalation of steam* from the croup-kettle be continued. Also, in any case, he finds that it is usually best to keep the air of the room moist, and at a temperature of 70° or 80°. The inner tube is taken out and cleaned as often as it becomes obstructed; but, on the second day, as a rule, he removes both tubes, cleaning them thoroughly and applying new tapes; while at the same time, also, the wound is cleaned, and the larynx examined as to its permeability. When returning the tube, he usually places a washer, formed of several folds of muslin, or other material, and smeared with a mild ointment, between the plate and wound, thus affording much comfort to the patient. The tube is permanently removed as soon as respiration can be carried on through the larynx.

DR. J. R. LEAMING recommends the following as often rendering excellent service :

℞. Potass. chlorat. . . . . ℥ ii.
Ammon. chlorid. . . . . ℥ i.
Aquæ cinnamom. . . . ℥ iii.
Syrup. senegæ . . . . ℥ ss.
Spts. æther. nitros. . . . ℥ ss.
Ext. glycyrrh. . . . . ℥ iss.

M. Sig. A teaspoonful to a tablespoonful, accord-ing to age, every two hours.

He also combines iron with the above, when required, according to indications.

DR. J. LEWIS SMITH directs that the atmosphere
of the room in which the child is placed, be constantly loaded with moisture, thus rendering the
cough looser and promoting expectoration. Any .
degree of heat, however, which would add materially
to the discomfort of the patient, must be avoided.
A temperature of 75° to 80° is usually required.
The following is, in his experience, the most efficient
solvent for the false membrane, and should be used
almost constantly, by means of the atomizer:

R. Liq. sodæ (or potassæ) . . ℥ i.
Aquæ calcis . . . . ℥ vii.
M.

This is freely sent directly to the inflamed surface
in the form of a spray.

For internal treatment, an emetic is first administered, for which purpose, as a rule, he prefers the
*yellow sulphate of mercury, in powder, in doses of
gr. ii.* In addition to this, he prescribes the following mixture with much benefit :

R. Potass. chlorat. . . . .
Ammon. muriat. . . āā ℈ i–ii.
Syrup. simp. . . . . ℥ i.
Aquæ . . . . . ℥ iii.
M Sig A teaspoonful every hour

As regards local measures, DR. SMITH employs *cold water*, especially in the early stages, preferring this measure, in most cases, to the use of poultices. This may be dropped constantly from a sponge, upon a compress laid over the throat of the child; or, two or three thicknesses of muslin soaked with *camphorated oil*, are applied over the larynx, so as to cover the neck in front, and over this is placed a bladder containing pieces of ice, or ice surrounded by oil-silk. If dyspnœa becomes severe, the *inhalation of oxygen* is often found of great service. As to the advisability of tracheotomy, he considers this resort, when indicated, proper and justifiable, and advises that it should not be delayed.

To control the inflammation DR. FRANCIS DELA-FIELD employs *leeches*, in strong, healthy children, applied over the region of the larynx; and following this measure with hot or cold applications around the throat, either *ice-bags*, or sponges wrung out of hot or cold water. The child is also made to breathe vapor of some kind, either-hydrating lime, or hot or medicated water; or the room is filled with steam, or the *steam tent* used. For internal medication, at the commencement he gives an emetic, usually *antimony in small doses*, combined with a small amount of opium. Regarding the administration of *calomel* to the point of salivation, he discountenances this practice, but advises that if given in small doses until the bowels move, and then discontinued, much benefit is often derived from its use. On this plan he prescribes:

 ℞. Hydrarg. chlor. mit. . . gr. ss-i.
 Sig. Dose, every one or two hours.

For the restlessness, small doses of *opium* in some form are given. When signs of insufficient aeration of blood are present, the inhalation of oxygen is employed, affording great relief. If, however, the disease is so far advanced that the breathing is obstructed, he resorts to *tracheotomy.* In regard to this procedure, DR. DELAFIELD advises that if the operation is performed while the local inflammation is running its course, the chances of recovery are much greater. Therefore, in cases where the bronchi are not involved, as soon as it is evident that the disease is not yielding, and deficient aeration is prominent, he operates at once, with considerable hope of this resort being not only beneficial but successful.

If seen at the onset of the disease, DR. A. HADDEN usually administers a dose of *castor oil*, sufficient to move the bowels. He then places a napkin wrung out of cold water about the neck, directs that the child be confined to bed, and prescribes the following:

℞. Tinct. ferri chlor. . . .

    Sodæ chlor. . . . āā ʒ i.

    Glycerinæ . . . . ℥ ss.

    Aquæ pur. . . . . ℥ iiss.

M. Sig. A teaspoonful every hour, to a child of three to six years.

In many cases, where the false membrane is confined to the tonsils, he applies the following astringent mixture to the affected parts, finding it of great value, and particularly so when the membrane does not extend to the larynx:

℞. Liq. ferri sulphatis . . . ʒ i.

    Glycerinæ . . . . . ℥ i.

M. Sig. To be applied repeatedly, by means of the throat brush.

Should the appearance of the membrane increase and with a tendency to spread, he often uses the following with good effect:

℞. Potass. chlorat. . . . . ʒ ii.

    Aquæ . . . . . . Oi.

M. Sig. Gargle.

When, however, the croupy condition becomes more manifest, with hoarseness, etc., he directs that the temperature of the room be maintained at 70° to 80°, and the air kept constantly moist witn steam. *Lime* is also slaked in the apartment so that the pa tient can breathe the fumes; this is found to have a

very soothing effect, and constitutes a means of relief upon which DR. HADDEN places great reliance.
In addition to these measures he frequently prescribes *bromine*, to be taken internally, with much
benefit. In some instances turpeth mineral is given,
as an emetic and for its alterative action. If the
symptoms increase in severity, in spite of these means
of treatment, and the croupy cough continues, he
often finds the application of a strong solution of
silver nitrate, by means of a probang, very serviceable in lessening the symptoms and checking the exudation.

If, however, all endeavors are unavailing and
the case still progresses unfavorably, he resorts to
tracheotomy. Regarding this procedure, he believes
that the operation, when indicated, is not only practicable but imperative, and in many cases the only
expedient that can hold out the least hope of saving
life. And in all cases he operates as soon as suffocation threatens, and it is evident that medicinal
measures are not likely to afford relief. After the
canula is in position, DR. HADDEN also considers it
a very essential part of the treatment, to make applications through the tube, by means of a soft
feather, of the above solution of iron and glycerine,
to the inner surface of the trachea. This he repeats
as frequently as every two or three hours during the

day, and believes the success following the operation to be, in a great measure, dependent upon this procedure.

DR. A. L. LOOMIS directs that the child be placed in a large room, the air of which is kept moistened, and the temperature at about 76° ; this he considers to be of the highest importance. In many instances the tent is used, and *steam from boiling molasses and water* carried into it. Lime vapor is also used with good effect. If seen at the onset, before the formation of any false membrane, he often derives much benefit from the administration of *quinine in large doses*, with the object of aborting the laryngeal inflammation, or preventing its assuming a croupous form. For this purpose he usually prescribes :

℞. Quiniæ sulph. . . . gr.xxx.

Sig. gr. v. every four hours, for twenty-four hours, to a child of three years.

When the false membrane is present, however, inhalation constitutes his sole reliance for safety. Should signs of imperfect aeration appear, *oxygen* is often of great service ; a stream being passed into the tent, or direct inhalation employed. In addition to these measures, hot cloths or sponges, squeezed dry (from boiling water), are continuously applied to the neck. For an emetic, when indicated, he prefers *zinc sulphate.* Direct local applications, he advises, do more harm than good.   He has also very little faith in the various specifics used in this disease. Regarding diet, this is made of the most nutritious character throughout, and when called for *stimulants* are freely given to sustain the vital powers.

Finally, DR. LOOMIS advises that the resort to tracheotomy, to be of service, must be made early, if at all ; and not, as is so often the case, after the child has passed beyond all hopes of recovery.

DR. AUSTIN FLINT believes that the operation of *tracheotomy* is not only justifiable, but that it is a duty which should, if possible, be performed, whenever it is evident that the child is dying from suffocation.

SPURIOUS OR CATARRHAL CROUP.
(ACUTE CATARRHAL LARYNGITIS.)

DR. J. LEWIS SMITH finds that most cases do well under suitable hygienic treatment, and without the use of any medicines whatever. He uses demulcent drinks, however, with much benefit, together with an occasional laxative. To relieve the cough, when troublesome, he uses a mixture of *paregoric and syrup of ipecac*, with marked benefit. Or, he sometimes finds a small *Dover's powder* fulfills this indication very effectually. For the restlessness a warm mustard foot-bath usually affords relief. *Inhalations* of a spray of glycerine and water are also used by him with good effect, together with mild applications, such as camphorated oil, rubbed over the part several times daily.

DR. FRANCIS DELAFIELD advises against any unnecessary alarm, thereby causing the child to be treated much more energetically than necessary. Mild cases, he finds, may require little or no treatment at all. To increase the secretion of mucus, and thus relieve the stridulous cough and breathing, he administers tartar emetic in small and frequently repeated doses, thus:

℞. Vini antimonii   .   .   . gtt. v–x.

Sig. Dose, according to age, every one to three hours.

If this amount causes vomiting, he directs that the drug must not be discontinued, but the dose diminished to gtt. v.; and if emesis is still produced, he then reduces this to gtt. ii–iii., or until he ascertains what amount the patient can take without vomiting. For the restlessness, one of the *opium* preparations is given in small doses; using either

Dover's powder, paregoric, or laudanum, or he often finds the *syrup of poppies* very serviceable. To relieve the dyspnœa, especially in spasmodic attacks, if it is very severe and the fever high, he gives an active emetic, either of the following being preferred for this purpose :

℞.   Hydrarg. sulph. flav.   .   gr. iii–v.

℞.   Vini antimonii   .   .   .   ʒ i.

℞.   Vini ipecac.   .   .   .   .   ʒ i.

In severe cases of dyspnœa, *tracheotomy* is indicated, and is, he finds, sometimes the only relief. In regard to this resort, DR. DELAFIELD advises not to delay too long ; after one or two severe attacks the third should not be waited for, as it may prove fatal. In those cases where the disease is protracted, or evinces a tendency to become so, lasting from ten days to two weeks, he considers the administration of a brisk purge, followed by tonics afterward, as the best means of breaking up the disease. On this principle, he gives *calomel every hour* until the bowels move, up to gr. v. being usually sufficient ; after which, he stops the calomel and administers quinine in doses of gr. $\frac{1}{4}$–i. By this means, satisfactory results generally follow.

DR. A. L. LOOMIS considers rest to the part of primary importance. He directs that the patient be confined in a warm room, the air of which is kept moist by steam, and the temperature at about 758. Where there is very little or no induration present, he finds that the persistent use of *inhalations of vapor* are of the greatest service, and believes this measure to be far superior to any other local means of treatment. He objects to the use of antimony or calomel, as well as to the application of blisters or leeches. For internal medication, he resorts to the early administration of *quinine in large doses*, with a view to rapidly producing the effects of the drug. On this plan, especially if begun at the outset, he has frequently succeeded not only in controlling but in arresting the progress of the disease. This, in a severe case, he usually gives as follows :

℞. Quiniæ sulph.   .   .   .   gr. xx.

Sig. This amount during the first twenty-four hours, to a child of three years.

Should these means fail and œdema come on, *scarification* is practiced. When, however, this cannot be done, or is useless, and suffocation is evident, he resorts to tracheotomy ; advising, moreover, against too great a delay in this respect.

LARYNGISMUS STRIDULUS.

(SPASMODIC CROUP.)

In ordinary cases, to relieve the spasm, DR. J. LEWIS SMITH places the child in a *warm bath*, of a temperature of 100°, as soon as possible after the onset of the attack. The bath is continued for ten or fifteen minutes, after which he administers an emetic, as follows:

℞. Syrup. ipecac. . . . . ℥ i.
  Sig. Dose, to a child under three years.

This is repeated in twenty minutes, or until vomiting occurs. Or, in the majority of cases, he finds the following combination more prompt in action than ipecac alone :

℞. Aluminis . . . . . ʒ ii.
  Syrup. ipecac. . . . .. ℥ i.
M. Sig. A teaspoonful.

In children over three years of age, he considers the *syrup. scillæ co. in doses of* ℥ *i* as having the best effect ; one or two doses being usually sufficient for this purpose. In addition to these means of treatment, if the bowels are not already loose, a purge is administered. Inhalations of steam are also used, and often the application of a *sinapism* over the neck is followed by marked benefit. Where the nervous character is more prominent, he derives much satisfaction from the administration of quinine, as follows:

℞. Quiniæ sulph. . . . gr. i.

Sig. Dose, three or four times daily, to a child of three to five years.

If cachexia is more or less marked, *tonics, iron,* etc., and plenty of fresh air, are required. As a rule, under this treatment, and with proper management, the symptoms disappear and convalescence soon follows.

DR. A. L. LOOMIS directs his attention primarily to the removal of the cause ; *e. g.,* indigestion, teething, etc. If the spasm continues he gives an emetic, or orders a hot bath, which are often found to act beneficially. In very severe attacks, however, when relief does not follow, and death seems impending, tracheotomy is resorted to.

DR. W. H. THOMSON finds the *cold douche* applied to the nape of the neck, often very serviceable in relieving the crowing respiration.

DR. A. A. SMITH advises that in most cases of spasmodic croup of reflex origin, the following will usually afford relief :

℞. Atrophiæ sulph. . . . gr. $\frac{1}{100}$.

Aquæ . . . . . Oss.

M. Sig. A teaspoonful every hour.

Or, if necessary, the dose is repeated every half-hour, according to the severity of the attack. Should the child's face begin to flush and show signs of the the physiological action of the drug, he then reduces the frequency of administration. Attention is also directed to the stomach, and if it contains anything which can be causing the spasm, an emetic is given; or a cathartic, if there is reason to suspect intestinal disturbance as the cause.

DR. FRANCIS DELAFIELD, in treating those cases which occur among weak and poorly nourished children, and in rickety subjects, directs his measures toward improving the general health and nutrition of the patient by means of *good food* and *tonics*, such as iron, quinine, cod liver oil, etc., and by sponging off the child with cold water each morning. Antispasmodics are avoided. At the onset of the attack he advises that nothing special in the way of treatment is called for; but if the spasm continues and suffocation is imminent, tracheotomy should be performed, although probably with very little benefit. In sudden emergencies, however, he directs that an ordinary *silver catheter* be passed through the opening of the glottis into the larynx and trachea, thus establishing respiration; and after this is again secured, the instrument may be withdrawn when the breathing will usually continue.

In other cases, where the affection is purely mus-

cular, DR. DELAFIELD administers an emetic, preferring the following :

℞.   Hydrarg. sulph. flav. .   .   gr. iii.–v.
    Sig. Dose, to be repeated if necessary.

Or he sometimes uses the *wine of ipecac in doses of*
℥ *i.* In either case he combines the use of the *hot
bath* very effectually with the above. After emesis
has taken place, should the child again commence to
have difficult breathing, the emetic is repeated. To
prevent a recurrence of the disease, he finds the
bromides and *assafœtida* most efficacious ; or sometimes opium, chloroform, or compound spirit of
ether have good effect.

HOOPING COUGH.

DR. J. LEWIS SMITH has treated several cases at
the New York Foundling Asylum, with very favorable results, by inhalation of the following mixture :

℞.   Acid. carbol.   .   .   .   .   ℨ ss.
    Potass. chlorat. .   .   .   .   ℨ ii.
    Glycerinæ .   .   .   .   .   ℥ ii.
    Aquæ   .   .   .   .   .   ℥ vi.
M. Sig. Spray.
6

This is to be inhaled three times daily, and from two
to five minutes each time. The good effects, he advises, seem at times to be due principally to the carbolic acid, although in one or two cases, where the
experiment was made of omitting the chlorate of
potash temporarily, the patient apparently did better
without it.

DR. A. JACOBI advises the importance of never
allowing the disease to run on for months, reasoning
that it is self-limited and must take a certain course.
But every endeavor should be exerted to cure the
case in five or six weeks, for if permitted to continue
it may give rise to secondary pneumonia. These
pneumonias, occurring secondary to hooping cough,
are most dangerous complications, and every means
should be employed to prevent them.

At St. Luke's Hospital, DR. J. R. LEAMING has
used *cerium oxalate in doses of gr. iii.–v.* with much
benefit in allaying the cough.

DR. H. S. DESSAU prescribes the following with
excellent effect, in the convulsive stage:

    ℞.   Chloral. hydrat. .   .   .   .   ℥ i.

        Potass. bromid. .   .   .   .   ℥ ii.

        Syrup. pruni virg.   .   .   .

        Aquæ   .   .   :   .   .   āā ℥ i.

    M. Sig. A teaspoonful three times daily, to a
child under one year.

DR. A. L. LOOMIS directs that all exposure be avoided, the child placed on a simple nourishing diet, and the condition of the alimentary tract be carefully attended to. In older children, he advises that they be instructed to restrain the cough when possible. To lessen the intensity of the paroxysms, he finds *belladonna* very serviceable in some instances; in others, hydrocyanic acid or cannabis indica affords most relief. He advises, however, that these drugs should be given in very small doses, and their effects carefully watched. Topical applications and *counter-irritation* he believes to be harmful; and he has no faith in the numerous specifics recommended for this affection. For the complications occurring in the course of the disease, these are promptly met in the usual manner, together with supporting measures. In many cases tonics, *iron*, *quinine*, and *cod liver oil*, are found very beneficial; and in all instances during convalescence, the administration of these remedies is usually called for.

The following mixtures are much used at BELLE-VUE HOSPITAL, in this disease:

℞. Potass. bromidi . . . gr. xvi.
  Syrup. ipecac. . . .
  Tinct. opii camph. . . .
  Syrup. lactucarii . . .
  Syrup. tolutan. . . . āā ℥ss.

Aquæ q. s. ad   .   ℥ iv.

M. Sig. A teaspoonful.

Or this:

℞. Tinct. nucis vom.   .   .   ℨ ii.

Vini ipecac.   .   .   .   ℨ iiss.

Syrup. sarsap. co.   .   .

Syrup. senegæ   .   .   .   āā ℥ iss.

M. Dose: A teaspoonful.

At the INFANTS' HOSPITAL the following is used with very gratifying results:

℞. Acid. nitric. dil.   .   .   .   ℨ i.

Syrup. pruni virg.   .   .   .   ℥ ss.

Aquæ q. s. ad   .   .   .   ℥ ii.

M. Dose: A teaspoonful.

EPISTAXIS.

DR. A. JACOBI advises that in all cases when the bleeding is profuse, it should be stopped at once, if possible, on account of the drain on the general system. He considers it a very bad plan to wash out the nares freely with water and weak astringents, since this interferes with coagulation which is essential for the control of the hemorrhage. *Styptic cot-*

ton, or the *persulphate of iron*, he finds very service-able in these instances. Or, in some cases, when necessary he uses *Belloc's canula*, closing up the entire nasal tract to effectually stop the bleeding. In addition, he advises that it is desirable to avoid all tight pressure about the body, and directs that the child be made to take very deep inspirations, filling the chest to the utmost extent, so that there may be as small an amount of blood as possible for the nose.

DR. G. M. LEFFERTS uses the following with very good effect:

R. Acidi tannic.   .   .   .   .   ℥ ss.

Pulv. acaciæ   .   .   .   .

Sacchari alb.   .   .   .   āā ℥ ii.

M. Sig. For insufflation.

DR. W. H. DRAPER finds that stuffing the nares with *lint dipped in ice-water* is often successful in arresting the hemorrhage. Or, in some instances, plugging the nares, either anterior or posterior, or both, is required. Internally, in cases of severe bleeding, he gives *ext. ergot. fl. in doses of m. xv–℥ ss.* every two hours, together with *gallic acid gr. ss–ii. in capsules*.

DR. ALONZO CLARK sometimes uses *pressure* very serviceably in checking the hemorrhage. When, however, the bleeding is very profuse, causing considerable pallor, etc., and in obstinate cases where

this means fails to relieve, he finds that a spray of
the following solution, when thrown into the nostril,
is quite efficacious:

℞. Liq. ferri persulph,   .   .   .   ʒ i.
    Aquæ   ⌒   . ʒ iv–vi.
M

PART III.

DISEASES OF THE DIGESTIVE ORGANS.

STOMATITIS.

In cases of simple stomatitis, DR. J. LEWIS SMITH considers the adoption of suitable hygienic measures of the greatest importance, and in many cases not any medicinal treatment whatever is called for. Locally he uses borax with very good effect, as follows :

R̶. Sodii biborat. . . . . ℨ i.

    Glycerinæ . . . . .

    Aquæ . . . . āā ℥ ss.

M.

The ulcerative form, he advises, requires the most careful attention to *cleanliness*, good surroundings, and plenty of fresh air. Tonics are also used, *iron* and the *vegetable bitters*, with also *cod liver oil*, when indicated. For local treatment, he considers the

chlorate of potassium as probably the most efficient
remedy, giving it thus:

R. Potass. chlorat. . . . ꝫ–ssi.
    Mellis . . . . . ℥ ss.
    Aquæ . . . . . ℥ ii.
M. Sig. A teaspoonful every two hours.

This, he directs, should be taken into the mouth
and permitted to flow over the affected parts, to obtain its local effect.

In cases of cancrum oris, DR. SMITH endeavors
to build up the constitution by good nutritious food,
tonics, etc. To arrest the gangrene, he has in several instances derived good results from the following:

R. Cupri sulph. . . . . ꝫ ii.
    Pulv. cinchonæ . . . . ℥ ss.
    Aquæ . . . . . ℥ iv.
M.

This he applies twice daily over the entire ulcerated
surface. If this means fails, however, and the gangrene
continues to spread, he then uses *strong muriatic
acid*, applying it carefully by means of a camel's-hair
brush, touching only the diseased parts. This is at
once followed by an alkaline wash.

DR. J. H. RIPLEY, and also DR. C. J. MACGUIRE,
has used *bismuth* successfully in a large number of
cases of cancrum oris, as well as in ulcerative stoma-

titis. The following plan is adopted by him. After
thoroughly cleansing the mouth and the cavity in
the cheek with the following disinfectant lotion :

℞. Potass. permang.　　.　.　.　　3 i.

　Aquæ .　.　.　.　.　.　　℥ iv.

M. Sig.　To be applied once daily.

he then clips away all the gangrenous tissue, and
packs the ulcerated surface with bismuth subnitrate,
repeating the process every three hours ; at the same
time washing out the mouth with a solution of carbolic acid, or the permanganate of potash solution.
The effect of this treatment is a clearing away of
the slough, frequently within twenty-four hours, a
lessening of the fetor, and a diminution of the general symptoms.　Constitutional treatment is also
carefully attended to.　*Syrupi ferri iodidi*, or quinine
and iron, and *cod liver oil* are administered internally,
together with a generous and nourishing diet.　This
local application of subnitrate of bismuth is claimed
to be almost a specific for this terrible malady.　In
nearly every instance the treatment has invariably
been followed by good results ; an immediate improvement in the appearance of the ulcers takes
place, and eventually a complete cure.

The value of *bismuth* in gangrenous stomatitis, as
well as in the ulcerative form, has also been fairly
tested at the NURSERY AND CHILD'S HOSPITAL.

The parts are first thoroughly cleansed with carbolic acid solution, cauterized with the *solid stick of silver nitrate*, and then packed with bismuth. *Iron, quinine, and whiskey*, are also given according to indications. In one case, however, bismuth failed, while alum cured the disease. From the results of the bismuth treatment in this hospital, and from experience, more confidence is expressed in the use of the solid nitrate of silver, followed by the application of alum, or, what is considered better, *alum combined with bismuth*, than in the use of bismuth subnitrate alone. Moreover, bismuth is regarded as being no more of a specific for ulcerative stomatitis and cancrum oris, than is pilocarpine for diphtheria.

TONSILLITIS.

In cases seen at the onset, DR. A. JACOBI has found usually prompt improvement take place under the administration of *potassium chlorate* and the *muriated tincture of iron*. Should any whitish films appear, after removing them he often uses a *spray of silver nitrate*, of the strength of 1–500, with good effect. In regard to unilateral inflammation of the tonsil or pharynx, he calls attention to the rarity of its occurrence, and advises that when present it should arouse a suspicion as to the existence of a contagious disease, and especially diphtheria.

DR. J. R. LEAMING finds that, if seen at the commencement of the attack, the disease may often be aborted. For this purpose he gives the following very successfully :

℞. Hydrarg. chlor. mit. . . . gr. ii.
  Antim. et potass. tart. . . . gr. ⅙.
  Sacchar. . . . . gr. iii.

M. Sig. Dose, every three hours, to a child of eight to ten years.

This he places dry upon the back of the tongue, and repeats the dose, as above, until free catharsis is produced. Or, when the inflammation is too far advanced to be aborted, he also uses this prescription with marked benefit in hastening the suppurative stage.

DR. FRANCIS DELAFIELD first administers a purge of *calomel*, after which he endeavors to abort the disease, if possible, by means of *cold* both within and without, together with the internal use of *aconite*. When the inflammation has gone on to suppuration, he then discontinues the aconite and resorts to warm applications, *steam inhalations*, etc. If there is much pain present, opium in some form is given to relieve it. When accumulations of mucus take place in the mouth, he finds a mild *solution of alum or borax* very effective in removing them, used either as a wash or gargle. In regard to *surgical interference*, he advises that if the swelling of the tonsils is not very great, and the discomfort of the patient not beyond endurance, the knife should not be resorted to. In any case, however, no attempt should be made to open the abscess before the fifth or sixth day; not until it has pointed, or until the yellow pus can be seen through the mucous membrane. As a rule, too early cutting, he finds, does harm, and while the sufferings of the patient must be the guide to this procedure, yet he considers it much better to let the tonsil break of itself, if possible. Where the uvula is involved, causing much distress, he obtains considerable relief from touching it with the scarifi-cator; or, if necessary, passing a bistoury into it, cutting from above downward. The following is very extensively employed at BELLEVUE HOS-PITAL :

R. Sodii biboratis . . . . 3 ii.
Fermenti . . . . .
Mellis . . . . . . āā ℥ ss.
Aquæ q. s. ad . . . . ℥ viii.
M. Sig. Gargle.

DR. F. H. BOSWORTH believes that he has, in many cases of commencing tonsillitis, aborted the disease before the suppurative period has been reached. To accomplish this, he gives *quiniæ sulph. gr. x.* followed by:

℞. Tinct. aconit. rad. . . gtt. v–x.
Aquæ . . . . . q. s.
M. Sig. Dose.

This is repeated until dryness of the throat, nausea, and ringing in the ears, show their constitutional effects.

DR. W. H. DRAPER is sometimes able to abort the inflammation by the use of *small blisters* and *poultices.* *Sinac,* which has been said to be a cure, will, in his experience, fail more often than it will succeed. Astringent washes and gargles are also used by him, at times, together with the internal administration of *aconite and belladonna* to control the circulation in this region. The following carbolic throat spray, used at NEW YORK HOSPITAL, is a most excellent and valuable combination :

℞. Sodii bicarb. . . . .
Sodii biborat. . . . . āā ʒ i.
Acidi carbol. . . . . gr. xl.
Glycerinæ . . . . ʒ vii.
Aquæ . . . . . ℥ viii.
M.

DR. F. A. BURRALL finds that gargles are of
marked service as a temporary substitute for poultices, a *warm gargle* frequently repeated affording
much relief and aiding suppuration.   He also advises that constitutional treatment should not be
neglected; a mercurial purge at the beginning is
considered of great value.   When the case is seen at
the onset, he prescribes the following with very
satisfactory result, and advises that if used at the
first intimation of the attack, its beneficial effects are
often remarkable :

    ℞. Acid. carbol. .   .   .   .   gr. xx.
        Glycerinæ   .   .   .   .   ℥ i.
        Sodii chloridi   .   .   .   ℨ i.
        Aquæ ferv.   .   .   .   .   Oss.
    M. Sig. Gargle, to be used every half-hour.

    DR. G. M. LEFFERTS directs attention to the fact
that when once a patient has suffered from this disease, a second attack is more liable to occur.   In
order to prevent this recurrence, if there is chronic,
inflammation he advises that the tonsil be excised ;
or if it is too small for this operation to be performed at any other time except when it is acutely
inflamed, it should be done then.   By this means he
believes that the chances of recurrence are almost
avoided, although even this, he finds, will not always
prevent a return of the inflammation.   For the

affection itself, when the development of suppuration is once assured, he treats it as an abscess elsewhere. He is doubtful as to the possibility of aborting it. *Silver nitrate*, when applied for this purpose, he finds, only produces separation of the slough, and causes pain and annoyance to the patient. Constant application of *heat and moisture*, internally and externally, he considers the best remedial means. As to medicated sprays, although possibly of some benefit, yet he thinks it questionable whether it is not merely the warm vapor that produces the result. He believes there is no better treatment than by the *hourly inhalation of steam*, as hot as can be borne, combined with the following gargle :

 ℞. Zinci chloridi . . . gr. ii–iii.

  Aquæ . . . . . ℥ i.

M.

Regarding gargles, however, he advises that they are not always of service, and sometimes do positive harm. Using them three or four times daily is considered insufficient. He directs the patient to use only a tablespoonful at a time, but to repeat it at least a dozen times a day. The partial act of swallowing he finds an excellent manner of doing this; or by throwing the head backward and simply letting the fluid flow back by force of gravity, is just as efficacious and more easily accomplished. He also objects to the application of *leeches* and *counter-irri-tants*, as these measures only serve to exhaust the patient's strength. In his experience, also, as little medicine as possible should be given, and if an emetic is tried, a non-depressant one should be used. *Cathartics* are also employed to relieve congestion of the alimentary tract. Regarding *scarification*, he believes there can be no doubt as to its propriety, since it causes little or no pain, affords relief at once, and may open into an abscess or make its sub-sequent spontaneous opening more easy by leaving a path for it. When the existence of pus is made out, he opens it at once. The following is also much used by him :

R.   Potass. chlorat.   .   .   .   ℨ ss–ii.
     Glycerinæ .   .   .   .   .   ℨ ii.
     Aquæ   .   .   .   .   .   .   ℥ x.
M.   Sig. Gargle.

In cases of chronic enlargement of the tonsils, DR. LEFFERTS thinks that too little regard is paid to active treatment, and directs attention to the unwil-lingness of parents to use remedies, as they believe that the children will outgrow it. He finds that where the case is comparatively recent, the tumor is soft and constitutional treatment will often result in its absorption. Local measures are also tried at times, such as application of *silver nitrate, massage* of the tonsil, etc. But when the enlargement is of long standing and the tonsils have become indurated, a train of more serious results may follow. In such cases, therefore, he resorts to *excision*, believing that the results of other methods, iodine, caustics, etc., are unsatisfactory. He advises, however, that it is not necessary to remove the whole gland, only the part in front of the anterior palatine fold; but the subsequent cicatrization must not be counted upon too much.

DR. BEVERLY ROBINSON also advises excision of the tonsils, where there is a condition of chronic hypertrophy and enlargement. If, however, this cannot be done, he sometimes uses silver nitrate with · good effect.

DR. FRANK H. HAMILTON regards an hypertrophied condition of the tonsils as one form of manifestation of the strumous diathesis. He finds, moreover, that these enlargements have usually their own definite period of growth and decline. They commence generally, according to his observations, about the second or third year of life, and usually reach their highest development about the tenth or twelfth year; after which they gradually diminish in size, and are rarely sufficiently large after about the twentieth year, to cause any inconvenience. Therefore he does not consider that the mere fact that the

7

tonsils are enlarged, justifies *excision*, and this he is
in the constant habit of saying to parents who bring
their children to him.   In any case, however, he
would not operate when the tonsils were in the
slightest degree inflamed, unless the danger of suffocation was imminent, from fear of setting up acute
inflammation or causing copious and frequently
recurring hemorrhage, either of which might prove
fatal.   He also considers the existence of a decided
hemorrhagic diathesis a strong contraindication to
operation.   DR. HAMILTON prefers to use his own
modification of Owen's instrument, after which the
bleeding is, as a rule, very trifling, and usually ceases
in a few moments of itself, or by washing the throat
with cold water.

PHARYNGITIS.

———

Where chronic pharyngitis, which, as a rule, is
more common in children than the acute form, is
associated with naso-pharyngeal catarrh, as is usually
the case, DR. A. JACOBI insists upon absolute cleanliness and *washing out the nares* regularly, several
times daily, with a tepid solution, using:

℞. Sodii chloridi . . . . ℥ i.

Aquæ . . . . . . Oi.

M. Sig. This amount once or twice daily.

This, he directs, must be snuffed up the nose until it can be spit out through the mouth. The plan is usually very successful, and he considers it preferable to the nasal douche, which sometimes causes bad results and is objectionable. In addition to this, a kettle of boiling water is kept constantly in the room, where the child can breathe it continually. Where a medicated application is desired, he employs the *nitrate of silver* in mild solution of gr. ¼–ii.—℥ i., according to the severity of the case. The stick or concentrated solution, however, he considers exceedingly dangerous, often setting up an incurable condition. As a rule, he uses:

℞. Argenti nitrat. . . . gr. ss–i.

Aquæ . . . . . ℥ i.

M.

This he injects twice a week, seeing that it enters the pharynx properly; while, during the intervals, the child is instructed to attend carefully to the salt and water washing.

DR. JACOBI also calls attention to an elongated condition of the uvula, which, when the child lies down, falls backward and touching the posterior wall of the pharynx causes a sudden tickling, thus

keeping up a continual cough.   This chronic state
of things in the pharynx, he advises, if allowed to go
on, results in dangerous consequences.   The cough
itself is a source of constant irritation, and eventually sets up a catarrh in the trachea and bronchi.
His treatment is to remove such an elongated uvula
as soon as possible.   If there is a choice, however,
he directs that it should not be done when diphtheria is prevailing.

DR. A. H. SMITH uses *silver nitrate* very effectually, in certain cases, where there is a condition of
naso-pharyngeal catarrh with muco-purulent discharge.   As a preparatory measure, however, he first
cleanses the parts thoroughly with a solution of sodium nitrate; after which he employs the following,
blown by means of the insufflator into the anterior
and posterior nares:

R. Argenti nitrat.   .   .   .   gr. v.
    Potass. sulphat. .   .   .   ℥ iss.
    Bismuth. subnit. ad.   .   .   ℥ i.
M. Ft. pulv.
Sig. To be used daily, or every other day.

DR. F. H. BOSWORTH cautions against permitting
children to habitually breathe through the mouth,
especially during sleep.   This he believes to be a
most prolific source of throat catarrh and bronchial
disorders.   In using inhalations, when the latter are

attended with much pain or irritation, he combines the following with evident satisfaction.

℞. Ext. hyoscyami . . . gr. v.

Aquæ . . . . . . ℥ i.

M.

DR. A. A. SMITH prescribes the following with marked effect, in those cases of pharyngitis occurring in older children, and associated with dyspepsia:

℞. Ammon. chlorid. . . . ʒ i.

Tinct. cubeb. . . . ℥ ss.

Tinct. gent. co. . . . . ℥ ss.

Glycerinæ ad . . . . ℥ iv.

M. Sig. A teaspoonful in water, every two or three hours.

As a rule, DR. FRANCIS DELAFIELD considers it best to begin with a purge, especially in pretty severe acute cases. For this purpose *calomel* is preferred by him, combined with a moderate amount of opium to relieve and quiet the patient. Following this, he administers:

℞. Tinct. aconit. rad. . . gtt. i–ii.

Liq. ammon. acetat. . . . ʒ i–ii.

M. Sig. Dose, every one or two hours, according to age.

The local measures adopted by him consist, at the commencement, in the application of *ice* to the throat, or the use of vapor or spray by the atomizer. The simple *vapor of steam* is used, or this is sometimes modified by the addition of vinegar to the water, tincture of cubebs, or other remedies. *Hot compresses* are also applied externally. After two or three days he resorts to astringents, in the form of *gargles and sprays of alum or borax* ; or zinc sulphate, tannic acid, and potassium chlorate are also used at times, either of which, he finds, act very well in the later stages.

DR. G. M. LEFFERTS considers the following very efficient when a sedative gargle is desired :

R. Potass. bromidi . . . 3 iss.
    Glycerinæ . . . . . 3 ii.
    Aquæ . . . . . ℥ x.
M.

When an astringent is called for, he finds the following very serviceable :

R. Aluminis . . . . . 3 i.
    Acid. tannic. . . . . . 3 i.
    Aquæ . . . . . ℥ x.
M. Sig. Gargle.

VOMITING—DYSPEPSIA

DR. J. LEWIS SMITH uses the following with marked benefit, in relieving the nausea accompanying intestinal trouble in infants:

℞. Acid. carbol. . . . . gtt.ii.

Aquæ calcis . . . . ℥ ii.

M. Sig. A teaspoonful.

This he directs to be taken in a teaspoonful of milk, using breast milk if the baby nurses, and repeated according to the nausea.

At the INFANTS' HOSPITAL the following is very extensively used:

℞. Bismuthi subcarb. . . .

Pepsini . . . . . āā gr.ii.

M. Sig. Dose.

DR. A. H. SMITH (also DR. G. B. FOWLER) has found beef peptone of excellent service in these cases. Where the infant refuses to nurse, followed by persistent vomiting, with probably diarrhœa, he gives *gtt. x. of beef peptone*, properly diluted, every two hours; at the same time stopping all other food and medication. This amount is, as a rule, gradually increased during the week. Its administration is almost invariably followed by improvement within a day or two; the vomiting and diarrhœa speedily and entirely cease, and the child soon regains its lost health and strength. Also in cases of dyspepsia of long standing, where the slightest indiscretion in regard to food is followed by a period of great suffering, and the usual remedies fail to give relief, beef peptone, in small doses at first and then gradually

increased to teaspoonful doses every three hours, affords more benefit and cuts short the attack quicker
than anything else. In fact, it is found to be uniformly successful in all these cases in improving nutrition, and in being assimilable when all other foods
are rejected. It may usually be given in water and
sweetened to the taste; but should the peptone be
objected to on account of the pronounced flavor of
meat, he often adds it to a simple broth with advantage, when it is generally taken very readily.

In those instances, so frequently met with, where
the child is nursing, and vomits or regurgitates its
food, DR. A. A. SMITH finds the following, in his experience, of excellent service in relieving this condition :

    ℞. Hydrarg. chlor. mit.   .   .   .   gr.i.
       Aquæ   .   .   .   .   . ˙   Oi.
    M.   Sig.   A teaspoonful every ten or fifteen min-
utes.

In preparing this, he advises that in order to dissolve it the calomel should first be added to aquæ
calcis ℨ i, and then to a pint of pure water. Or, when
the vomiting is accompanied with mucous discharge,
he gives the following with good effect :

    ℞. Hydrarg. chlor. corros.   .   .   gr.ss.
       Aquæ .   .   .   .   .   .   Oi.
    M.   Sig.   A teaspoonful every fifteen minutes.

In those cases of vomiting due to indigestion, commonly found in young children, he gives:

℞. Vini ipecac. . . . . gtt.i.

Aquæ . . . . . q. s.

M. Sig. Dose, every ten or fifteen minutes.

This he finds of the greatest benefit, and advises that it will often arrest the most obstinate vomiting, as well as any diarrhœa that may be present ; and, moreover, when administered in this manner, the drug is not in the least nauseous and is easily taken.

In other instances, associated with a simple non-inflammatory diarrhœa, DR. SMITH prescribes *hydrarg. cum cretæ, gr. $\frac{1}{24}$ every fifteen or twenty minutes.* This he recommends with much satisfaction in relieving this often exceedingly troublesome condition.

Many cases of nausea and vomiting in children are also promptly and thoroughly relieved by the use of *hot water.* This is found to be especially useful in cases where these symptoms are purely reflex, and in the colic of newly born infants.

DR. W. A. HAMMOND uses the following very successfully, in the dyspepsia occurring in older children and dependent upon nervous causes:

℞. Sodii bromid. . . . . ℥ i.

Pepsin. citrat. . . . .

Pulv. carbon. . . . . . āā ℨ iii.

Aquæ   .   .   .   .   .   ℥ iv.

M. Sig. A teaspoonful or less, according to age, three times daily.

DR. F. DELAFIELD prescribes the following, which is a favorite with him, in cases of dyspepsia in older children :

℞. Pepsini   .   .   .   .   gr. iii–v.
Pulv. cubeb.   .   .   .   gr. v–x.
Bismuthi subnit. .   .   .   gr. v–x.
M. Sig. Dose.

GASTRITIS.

In the acute catarrhal gastritis of infants, in moderately severe attacks, DR. F. DELAFIELD frequently finds that simply diminishing the food, and doing little or nothing else for twenty-four hours, will be all that is required. In severe cases, where there is high fever, he often employs sweating very serviceably ; placing the child in the *hot bath* for a few minutes and then wrapping in a hot blanket. Small pieces of cracked ice may also be given at intervals. In many instances, however, he finds that the fever is of short duration, and calls for no treatment whatever. To control the persistent vomiting, if the infant is brought up on the bottle, he gives *sodii bi-*

carbonas with milk; or if the child is nursing, he administers milk with the bicarbonate of sodium in addition to its ordinary food; in either case giving *sodii bicarb. gr. iii–iv. in ℥ ss. of cream and milk,* or at times using sugared water. In other cases he employs the following method of administration very effectually:

℞. Sodii bicarb. . . . .   ℥ i.

    Cremor. lactis . . . .

    Aquæ   . . . . . āā ℥ iv.

M. Sig. A teaspoonful, to be taken every hour whether retained or not.

Sometimes, however, he obtains excellent results from *calomel, in small doses of gr.* $\frac{1}{16}$–*ss.*

In children over three years of age he substitutes milk for the water in the mixture, and gives double the dose, feeding the child on this alone while the attack is going on. Or, often he finds that *brandy in small doses* is very serviceable. To relieve the pain, which is usually prominent in these older children, he applies *anodyne poultices,* or sometimes *turpentine stupes,* with good effect, and, as a rule, administers small doses of Dover's powder, just sufficient to diminish the pain. Where *morphia* is used for this purpose, he advises that it should be given by the stomach, usually beginning with m. ii.

ENTERALGIA (Colic)—ENTERITIS.

Dr. J. W. McLane prescribes the following with very gratifying effect in the colic of babies:

℞. Chloral. hydrat.   .   .   .
    Sodii bicarb. .   .   .   .
    Potass. bromid.   .   . āā gr. vi–viii.
    Aquæ rosæ   .   .   .   .   ℥ i.

M. Sig. A teaspoonful, repeated every half-hour as required.

In many cases, however, he prefers to use mint water—not peppermint, but green mint—in the above formula, instead of aqua rosæ.

Dr. F. A. Burrall recommends the following as an excellent carminative for infants :

℞. Tinct. valerian. ammon.   .   .   ℨ ii.
    Lactopeptin.   .   .   .   gr. xxxii.
    Sodii bicarb.   .   .   .   gr. xii.
    Glycerinæ   .   .   .   .   ℨ ii.
    Aquæ   .   .   .   .   .   ℨ vi.
    Aquæ aurant. flor.   .   .   .   ℥ i.

M. Sig. M. xx–lx. in ℨ i–ii. of warm water as needed.

This he has found to be an exceedingly valuable combination, where it is desirable to avoid an opiate.

As a rule, in cases of spasmodic colic, DR. AUSTIN FLINT pays little or no attention to the cause of the attack or to the presence of constipation ; but directs his treatment to the relief of the spasm, especially by the use of opiates, believing that so long as the spasm exists cathartics are of little or no avail.

DR. A. A. SMITH considers the following very efficacious in relieving an attack of colic depending on flatulence :

R. Spts. chloroformi . . . ℥ ss.

Tinct. cardam. co. . . . ℥ ii.

M. Sig. M. xv–xxx. in a tablespoonful of water, every half-hour until relieved.

DR. F. H. HAMILTON finds that opium does not always cure an attack of colic. His later experience has been that this affection is most quickly and most permanently relieved by a full dose of some aromatic and stimulating cathartic, such as *tincture of rhubarb with ginger.* In certain cases, however, he finds that only a full dose of some active sedative will succeed. But he places greatest reliance on the *mechanical effects of posture* (regarding displacement or doubling of the gut as the cause, at least in the majority of cases, rather than spasm or fecal obstruction), not, however, as a positive cure for all cases, nor as a substitute for any other suitable mode of treatment; but particularly as a supplement to other means, and which may sometimes prove effectual, or at least useful. In older children, he directs that the patient be made to assume a position elevating the hips with pillows, or over the end of a sofa. Occurring in infants, he advises that a similar position be secured, or that the child be raised by the feet as if in the act of applying a diaper; thus elevating the lower part of the body so as to cause the heavy organs, such as the liver and spleen, to fall toward the head, dragging the intestinal viscera with them. By this means, DR. HAMILTON often finds that there is an almost immediate discharge of gas from the rectum, and, in infants, often a free fecal evacuation, with the effect of prompt and complete relief of the colic.

At the INFANTS' HOSPITAL the following is constantly employed:

℞. Infusi anisi (℥ ii–Oi.)   .   .
    Genevæ.   .   .   .   .   āā ℥ i.
M. Dose: Half a teaspoonful.

At the NEW YORK HOSPITAL the following combination is considered to be highly valuable as a carminative:

℞. Tinct. opii .   .   .   .   gtt. xx.
    Ol. anisi   .   .   .   .
    Ol. caryophyl.   .   .   .
    Ol. gaulth. .   .   .   .   āā gtt. ii.
    Tinct. asafœtid. .   .   .   .   ℨ i.

Magnes. carbon. .   .   .   .   ℥ i.

Aquæ menth. pip.   .   .   .   ℥ iii.

M.

In cases of nursing infants, where after taking milk there follows a colicky condition, with griping pains, etc., the following mixture is found by many to afford speedy relief to these often frequent and exceedingly annoying attacks:

℞. Potass. carbon.   .   .   .   gr. ii.

Ol. cajuput.   .   .   .   .   m. i.

Aq. anethi   .   .   .   .   ℥ ii.

M. Sig. Dose, three or four times daily.

In pure enteritis. DR. F. DELAFIELD advises the importance of an early diagnosis, since the treatment varies greatly from that of peritonitis, with which it is sometimes apt to be confounded. He first administers a purge, usually preferring a full dose of castor oil; or *calomel, gr. ii–vi.* in three divided doses one hour apart, adding also a little Dover's powder to the calomel. After this, and during the first day or two of the disease, he finds the use of *opium* to relieve the pain, very advantageous, but directs that this must not be continued after the fever has much diminished, even though the pain persists, as opium then is apt to make matters worse. During this early period he advises that the *diet* be regulated as near to nothing as possible, usually consisting of

milk, until the fever has subsided. After the first
twenty-four or thirty-six hours, he places the patient
on small and gradually increasing doses of *ipecac;* at
this stage of the disease, also, he orders a diet of beef-
tea for a few days, followed after a while by the more
ordinary articles of food, avoiding starches and milk.

DIARRHŒA.

In infantile diarrhœa due to indigestion and attended by acidity, DR. J. LEWIS SMITH finds the
following very effective:

 ℞. Pulv. ipecac. . . . . gr. ss.
   Pulv. rhei   . . . . gr. ii.
   Sodii bicarb. . . . . gr. xii.
 M. Div. in chart. No. xii.

   Sig. One powder every four to six hours, to an
infant one year old.

In a large majority of cases, however, he employs
the following combination. If it fails to relieve, and
the regimen has been carefully attended to, he con-
cludes that in all probability there is inflammation of
the intestinal mucous membrane.

 ℞. Tinct. opii deodor.   . . gtt. xvi.
   Bismuthi subnit. ·.   . . . ℨ ii.

Syrupi simp. . . . . ℥ ss.

Mist. cretæ . . . . ℥ iss.

M. Sig. A teaspoonful every three or four hours, to an infant of one year.

For the simple diarrhœa occurring in older children he generally gives:

℞. Bismuthi subnit. . . . gr. xxx.

Cretæ præcipit. . . . gr. xxx.

Pulv. opii . . . . gr. i.

M. Div. in pulv. No. x.

Sig. One powder every three or four hours, as required. .

In giving *bismuth* to children, DR. E. G. JANEWAY calls attention to the mode of administration. He finds that it is very common for the physician to direct that it be taken in water or milk, and it is noticed by the mother or nurse that the drug sticks to the bottom of the spoon, and makes the child gag in swallowing it. Therefore, in young children, he advises that it is not best to give bismuth in large doses, and to administer it with something which will hold the powder in suspension; some mucilaginous material. He also directs that the mother be instructed as to the effect of the drug on the color of the passages, and to the disagreeable odor which is sometimes present from the fermentation of the

8

mucus; thus avoiding unnecessary and often serious alarm.

DR. A. A. SMITH recommends the following as very serviceable in cases of diarrhœa dependent upon intestinal irritation :

℞. Ol. ricini   .   .   .   .   gtt. v.
    Sacchar.   .   .   .   .   q. s.
M. Sig. Dose, to be given every two hours.

He also uses the *wine of ipecac, in doses of gtt. i. every ten or fifteen minutes*, with excellent effect in arresting diarrhœa due to indigestion; this he finds particularly useful where there is also vomiting present, and advises that when given in this manner, the drug is easily taken and has not the slightest nauseating effect. *Hydrarg. cum cretæ, gr. $\frac{1}{24}$ every fifteen minutes*, is also recommended by him in cases of simple diarrhœa associated with vomiting. In older children, where an anodyne and stimulant is indicated, he prescribes the following with much satisfaction :

℞. Spts. ammon. aromat.   .   .   .
    Spts. chloroformi .   .   .   .
    Tinct. camphoræ   .   .   .   .
    Tinct. opii deodor .   .   .   . āā ʒ ii
    Tinct. capsici   .   .   .   . ʒ i.
M. Sig. M. x–xx. in a wineglass of water.

In diarrhœa due to food, whether acting as a

purge or as an irritant, DR. F. DELAFIELD usually controls it by the use of *castor oil*, or sometimes rhubarb. Where, however, the irritation and passages continue, he administers *opium*, either alone or with bismuth or bicarbonate of sodium. In the diarrhœa of young children which commonly occurs in hot weather, he finds that the mildest of these cases hardly need any treatment at all; or he sometimes gives:

R. Ol. ricini . . . . ℨ i– ℥ ss.

Tinct. opii . . . . gtt. ii–iv.

M. Sig. Dose, according to age.

Should the attack continue, he then considers it unwise to use purgatives. The indication is to relieve the irritant condition of the intestine, to accomplish which he administers opium, as above, in very small doses, generally using Dover's powder or paregoric. This, also, he finds is best combined with small doses of ipecac, or rhubarb, thus:

R. Tinct. opii . . . . gtt. iii–iv.

Pulv. ipecac. . . . . gr. ⅛.

M. Sig. Dose, every three or four hours.

Or, *pulv. rhei, gr.* ¼ is given instead of the ipecac. He further directs that the child be kept quiet and in a cool place, avoiding fruits and feeding on milk and starches.

The following mixture is found by many to be a

valuable prescription in cases of non-inflammatory diarrhœa, where there is not much pain or tenesmus, and where the evacuations, though watery, are fecal and contain but little mucus:

℞. Magnes. sulph. . . . . ℥ i.
Tinct. rhei . . . . . ℥ ii.
Syrup. zingib. . . . . ℥ i.
Aquæ carui . . . . ℥ x.

M. Sig. A teaspoonful three times daily, to a child of one year.

At the INFANTS' HOSPITAL the following is very considerably employed:

℞. Bismuthi subcarb. . . . gr. ii.
Acidi tannici . . . . gr. i.
Pulv. ipecac. co. . . . gr. ¼.

M. Sig. Dose, repeated as necessary.

At the HART'S ISLAND HOSPITAL (child's), the following diarrhœa mixture is kept constantly on hand, and is considered to be a most excellent combination:

℞. Tinct. capsici . . . . ℨ i.
Tinct. catechu . . . .
Tinct. kino . . . .
Tinct. krameriæ . . . . āā ℨ iv.
Tinct. opii . . . . . ℨ iii.
Spts. menth. pip. . . . ℨ ii.

Spts. camphoræ .   ᛁ   •   •

Aquæ   •   •   •   •   •   āā ℥ iv.

M. Dose : M. xxx.

At BELLEVUE HOSPITAL, the following are used with much satisfaction :

℞. Tinct. opii camph.   •   •   •

     Syrup. rhei arom.   •   •   . āā ℥ ss.

     Aquæ calcis   •   •   •   •   ℥ ii.

M. Dose : A teaspoonful.

Or :

℞. Hydrarg. cum cretæ   •   •   gr. ¼.

     Pulv. bismuth. subnit.   •   •

     Pulv. pepsin.   .   •   •   . āā gr.iii.

M. Sig. Dose.

Also :

℞. Tinct. opii .   •   •   •   •

     Tinct. rhei arom.   •   •   •

     Spts. camphoræ   •   •   . āā ʒ ss.

     Tinct. cardam. co.   •   •   •   ʒ ii.

     Aquæ anisi q.s. ad   .   •   •   ℥ iv.

M. Dose : A teaspoonful.

For the administration of castor oil, the following combination is rendered exceedingly palatable, and is readily taken by children :

℞. Pulv. gum. acac.   •   ᛁ   •   ℥ i.

     Syrupi   •   •   •   •   •

     Glycerinæ .   •   •   •   . āā ℥ i.

Aquæ	.	.	.	.	℥ iii.
Ol. ricini	.	.	.	.	℥ vi.
Ext. vanillæ	.	.	.		
Spts. vini gall.	.	.	.	.	āā ℥ ii.
Ol. cinnam. ver.		.	.	.	m.v.

M. Dose: Double the quantity of oil intended to be given.

INTESTINAL CATARRH.

In cases of diarrhœa with intestinal catarrh in infants, when due to improper food which has not been digested and thus acts as an irritant, DR. A. JACOBI places the child on a diet of milk, diluted one-half or less with gum arabic water, or barley water. This dilution he considers very important in the treatment. Also, to prevent coagulation of the milk by the over-acid stomach, he adds an antacid ; avoiding the sodium and magnesium salts, on account of their purgative action, and giving *calcium carbonate in doses of gr. iii.* In some instances, and especially in bad cases occurring in the summer time, he finds that cow's milk is not digested at all. He therefore prohibits its use at once, and gives *barley water* alone, or substitutes the white of egg for the milk.

Or, if the diarrhœa is very marked, he forbids the milk altogether. This treatment, he advises, must be kept up for from one to two weeks, according to the previous duration of the catarrh. The diarrhœa, he directs, should be checked at once to avoid its injurious effects on the abdominal glands. If there is still any irritant substance in the intestine, he administers a dose of *castor oil, ℥ i. given in hot milk.* This is usually efficient.

For local measures, to act on the inflamed mucous membrane, he considers bismuth superior to other remedies, as it acts not only as a protection to the inflamed surface, but also, at the same time, as an anti-fermentative. He generally gives it in combination with opium, to check the undue amount of secretion and peristaltic action; thus :

℞. Bismuthi subnit. . . gr. ii–iii.

Pulv. Doveri. . . . . gr. ⅓.

M. Sig. Dose, every three hours.

Under this treatment, with careful attention to feeding, which he considers most important, the results are usually very satisfactory.

When the diarrhœa is associated with chronic catarrh dependent upon ulceration of the intestine, DR. JACOBI advises that treatment with astringents only is useless. In these cases the indications are to stop the diarrhœa, and heal the ulcerations. *Lime*

has the effect of neutralizing the abnormal, and perhaps the normal acids, and now and then he finds
that it does good; but if given for a long time it is
apt to cause obstinate constipation, hence care is
advised as to the amount used.   Bismuth, which
acts as an anti-fermentative, he considers very serviceable. In children of six to twelve months of age,
he gives:

℞. Bismuthi   .   .   .   . gr. ii–iii.

Opii   .   .   .   .   . g. $\frac{1}{60}$–$\frac{1}{80}$.

M. Sig. Dose, every three hours.

In some instances he also prescribes:

℞. Argenti nitrat.   .   .   . gr. $\frac{1}{60}$–$\frac{1}{40}$.

Aquæ   .   .   .   .   .   . ℥ i.

M. Sig. Dose, every two hours.

Care must be exercised about giving astringents,
as these young children must also take milk.

As regards the *diet*, this he believes to be a matter
of the greatest importance. In many cases where
milk cannot be digested, he advises that the patient
be made to eat the milk by spoonfuls, not to drink
it ; and to add to it a little common salt, as by this
means it is made more like mother's milk which
contains more sodium than cow's milk. When,
however, cow's milk cannot be tolerated at all, he
directs that it be mixed with some glutinous substance, so that it can be acted on very slowly by the

gastric acids; usually ordering one-third to one-tenth of cow's milk, after having boiled it, to be mixed with barley water, the latter also being boiled alone for twenty minutes. If cheesy matter is still found in the passages, he omits the milk altogether, and gives one or two *whites of egg with barley water* in twenty-four hours. On this plan, he generally administers about one-half pint of milk, three pints of barley water, one white of egg, and with this a teaspoonful of brandy, during the twenty-four hours.

In cases of protracted diarrhœa with chronic rectal catarrh, where prolapsus occurs after every act of defecation, and the mucous membrane is swollen with more or less hypertrophy of the part, DR. JACOBI finds the following ointment of rare service:

℞. Ext. nucis vomicæ . . pars i.

Adipis . . . . . part. xx.

M.

Finally, he calls attention to the fact that young children with catarrh of the colon and follicular colitis, are very apt to have a dilated colon; and in these dilated artificial diverticula masses of fecal matter are retained, the serum is absorbed, and in this way the hard feces form, resulting, after a time, in paralysis of the colon and perhaps of the rectum and sphincter.

DR. J. LEWIS SMITH prescribes the following with

much satisfaction, in the intestinal catarrh of infants where the nausea is extreme:

℞. Acidi carbol. . . . . gtt. ii.

Aquæ calcis . . . . ℥ ii.

M. Sig. A teaspoonful.

This he directs should be taken with a teaspoonful of milk, using breast milk if the baby nurses, and repeated as often as the nausea requires. Where the disease has passed into the chronic stage, he administers tonics, etc., and advises the most careful attention to feeding and hygiene. For a tonic in these cases he gives the following with marked benefit:

℞. Tinct. calumbæ . . . ℨ iii.

Liq. ferri nitratis . . . gtt. xxvii.

Syrup. simplic. . . . . ℥ iii.

M. Sig. A teaspoonful every four hours, to an infant of one year.

This he finds especially valuable where there is much pallor, loss of strength, and emaciation, showing failure of the vital powers.

In cases of diarrhœa where no history of weather influences exists, and with vomiting and considerable fever, DR. F. DELAFIELD gives sedatives at once, using opium either alone or with bismuth, ipecac, or sodii bicarb.; often prescribing the following:

℞. Pulv. Doveri . . . gr. ss–i.

Bismuthi subnit. . . gr. iii–iv.

M. Sig. Dose, every three or four hours.

Or, *ipecac, gr.* ⅛, or *rhei, gr.* ¼, is sometimes substi‑ tuted for the bismuth in the above combination. In some instances, however, where although the child improves, yet it does not entirely recover, and the disease becomes protracted, he advises that it is use‑ less to persist with the opium treatment after a moderate time. In these cases he directs that a *change of air* be obtained, which alone will often effect a cure. Where this is not sufficient, he then administers *hydrarg. cum cretæ, gr.ss. two or three times daily*, for three or four days, and usually with very marked benefit. Or, he frequently gives *calo‑ mel gr.* ⅛–*ss.* in the same manner.

Regarding *diet* in these protracted cases, if the child is nursing he sometimes allows it to continue, or may change the nurse, or try hand-feeding ; vary‑ ing his course with the indications. If the child is fed with the bottle he generally changes the food, sometimes adding other articles, where the diet has been milk alone, or, in other instances, directing that the child be nursed. In other children, he usually obtains good results from feeding on *raw meat*, beef‑ steak scraped to a pulp, with a little salt added, and keeping this up for some time (avoiding a too long continuance, however, from danger of scurvy).

In the severest forms of the disease, where although the attack begins in a mild way, yet in spite of treatment the diarrhœa continues, assuming a dysenteric character, and the stools finally become of a starchy consistency, a whitish color, and at the same time more and more abundant, while the constitutional symptoms are extremely marked and emaciation rapid, DR. DELAFIELD proceeds as follows: In the milder forms he generally finds it best to commence with castor oil, thus:

  ℞. Ol. ricini . . . . ℥ i–℥ ss.
    Tinct. opii . . . . gtt. ii–iv.
  M. Sig. Dose, according to age.

This, in some cases, may be sufficient. Should the movements continue for a time, he then administers *opium with bismuth*, as in other instances. If the disease still persists, the remedies which he finds most serviceable are the mineral acids, with opium ; more especially dilute hydrochloric acid, which he administers, at first, as follows:

  ℞. Acid. hydrochlor. dil. . gtt. x–xx.
    Tinct. opii . . . . gtt. i–ii.
  M. Sig. Dose, three or four times daily, according to age.

As improvement takes place, he diminishes the opium, keeping on with the acid, until the child is much better, when he stops the opium altogether, the acid, however, being still continued.

For the *diet*, in these cases, DR. DELAFIELD finds that these children are best fed with *cod liver oil and meat*, and with as little starch as possible; older children he places on a diet of cream and milk with meat or oil, while to very young children he gives cream and milk. If possible, he advises a *change of air*, as exceedingly beneficial, except in chronic cases. He also advises care in regard to the general management of the child, regulating the bath, clothing, etc.

DR. A. A. SMITH often gives the following with advantage, in those cases where the diarrhœa is accompanied by mucous passages, showing a slight degree of inflammation:

℞. Hydrarg. bichloridi . . gr. ¼.

Aquæ . . . . . ℥ viii.

M. Sig. A teaspoonful every hour.

Where a more intense inflammatory action is present, with straining, and the passages consist of a jelly looking matter, but not truly dysenteric, he prescribes the following very effectually:

℞. Ol. ricini . . . . gtt. v.

Sacchari . . . .

Mucilag. . . . .

Aquæ . . . . . āā q. s.

M. Sig. Dose, every hour.

This remedy, thus administered, he has found to render excellent service in many of these cases.

At BELLEVUE HOSPITAL the following form of combination for *Dover's powder* is very much used, and is considered preferable to the officinal powder:

℞. Pulv. opii (12 per cent.)   .

    Pulv. ipecac.   .   .   .   āā gr. i.

    Sacchari lactis   .   .   .   gr. viii.

M.

DYSENTERY.

In catarrhal dysentery, DR. FRANCIS DELAFIELD considers rest all important; no matter if the constitutional symptoms are slight, or even absent, still he insists on the necessity of *absolute rest.* In milder cases this may be all that is necessary, and the disease succumbs without further treatment. If seen early he usually begins with:

℞. Ol. ricini   .   .   .   .   ℨ i–iii.

    Tinct. opii   .   .   .   .   gtt. iii–v.

M. Sig. Dose.

Where, however, the pain and irritation about the rectum continue, opium is given to control it; either by the mouth in the form of *bismuth and opium powders,* or by the rectum in small enemata of starch and laudanum, or by suppositories. Regarding *diet*, this, he advises, needs careful regulation ; all solid food must be avoided and the child fed on milk, either alone or with starches, arrowroot, rice, etc. By this means, these cases usually terminate favorably.

In croupous dysentery, if the case is seen at the onset, and in those resembling the catarrhal form, he generally administers *ol. ricini and opium* in sufficient doses to afford quiet and relieve the pain. After the first dose he often finds it necessary to give only the opium. In the more severe forms, however, after beginning with castor oil and opium, he then administers *ipecac in large doses* every one, two, or three hours, according to the severity of the disease and the patient ; at the same time giving opium in small doses. In three or four days he repeats the castor oil, following it with ipecac as before, and then falls back on opium. During this treatment the ol. ricini may have to be repeated several times, depending upon the duration of the disease. Where the pulse becomes feeble, *stimulants* are given. After the inflammatory process has run its course, and ulcers are present, the castor oil and ipecac are discarded, and the *dilute mineral acids* resorted to ; or suppositories of *iodoform* are used with good effect. Stimulants are also employed at this stage, and the strength of the patient supported with good nourishment, etc., the diet consisting of arrowróot, rice and milk.

Where the disease tends to become protracted, DR. DELAFIELD advises especial care regarding rest and diet. A strictly milk food he finds particularly serviceable in the catarrhal form. Moderate stimulation is also employed, and a very gradual return made to the ordinary diet. For the general health, he administers *tonics, iron and quinine,* and advises a change of residence to a warm climate, together with plenty of fresh air.

As a rule, in most instances, DR. J. LEWIS SMITH finds that marked improvement is often derived from administering the following:

 ℞. Pulv. ipecac. co. . . . . ʒ i.

   Bismuth. subnit. . . . . ʒ ii.

 M. Div. in pulv. No. xxiv.

   Sig. One every two to four hours, to a child of five years.

In conjunction with this, he also gives:

 ℞. Tinct. opii deod. . . . ʒ ss.

   Bismuthi subnit. . . . ʒ ii.

   Aq. menth. pip. . . . .

   Syr. zingib. . . . . āā ʒ i.

 M. Sig. A teaspoonful, every two to four hours.

Besides this means of treatment, he directs that the most careful attention be paid to diet, etc.

DR. G. B. FOWLER (also DRS. A. H. SMITH and LORDLY) has employed beef peptone, in cases of acute dysentery, with remarkable effect. Under the use of this remedy the results have been exceedingly good; the distressing symptoms subside, the vomiting and stools are speedily controlled, nutrition is markedly improved, and the child soon begins to recover its health and strength. The administration is usually commenced with *gtt. x. of beef peptone every two hours*, and then gradually increased to doses of ℨ i. every two or three hours, to infants of six months and over. He generally gives it in water, sufficiently diluted, and sweetened to the taste. Should, however, the strong taste of meat be objected to, it is often added to a simple broth, and in this way readily taken. Moreover, as a diet, in these cases, he has found peptone to be assimilated when all other foods are rejected.

In older children, DR. ALONZO CLARK uses the following with much advantage :

R. Bismuth. subnit. . . . gr. v.

    Morphiæ sulph. . . . gr. $\frac{1}{18}$.

M. Sig. Dose, two or three times daily.

By this means he often obtains relief in these cases; advising, also, that the influence of proper diet and surroundings must receive careful attention.

9

At BELLEVUE HOSPITAL the following mixture is employed :

℞. Acid. nitric. .   .   .   .   m. viii.
    Tinct. opii   .   .   .   .   m. xl.
    Aquæ camphor.   .   .   .   ℥ viii.
M. Dose : A tablespoonful or less, according to age.

CHOLERA INFANTUM.

(SPURIOUS HYDROCEPHALUS.)

In the earlier stages of the disease, DR. J. LEWIS SMITH finds that the following is often of great service :

℞. Tinct. opii deod.   .   .   gtt. xvi.
    Bismuth. subnit.   .   .   .   ℨ ii.
    Syrupi   .   .   .   .   .   ℥ ss.
    Aquæ   .   .   .   .   .   ℥ iss.
M. Sig. A teaspoonful every two or three hours, to a child of one year.

When the vomiting and diarrhœa are excessive and peristent, so that no food is retained, he uses the following with marked benefit :

℞. Spts. ammon. aromat.   .   .   ℨ ss–i.
    Tinct. opii   .   .   .   .   gtt. xvi.

Bismuth. subnit. . . . ℥ ii.

Mistur. cretæ . . . . ℥ iss.

Syrupi simp. . . . ℥ ss.

M. Sig. A teaspoonful every two or three hours, to a child of eight to twelve months.

This he continues, as directed, until these symptoms are fully controlled. Stimulants are also administered as required. When the hydrocephaloid stage is reached, he gives the following with advantage :

℞. Tinct. calumbæ . . . ℥ ii.

Liq. ferri nitratis .. . gtt. xviii.

Syrupi . . . . . ℥ ii.

M. Sig. A teaspoonful.

On this plan, together with the most careful attention to diet, etc., his results are often very favorable.

At the NEW YORK FOUNDLING ASYLUM the *hrdrobromide of cinchonidia*, by hypodermic injection, has been employed in numerous instances of cholera infantum, and usually with the most satisfactory results. In many cases with high temperature, and profuse diarrhœa and vomiting, this remedy has afforded relief in a very short time. The injection, it is advised, should be repeated at least once, if not twice, in the twenty-four hours. A ten per cent. solution is generally employed, and small doses of *carbolic acid* given internally in conjunction, with also cold baths and stimulants as required. By this method, also, not only a speedy action of the remedy is obtained, but any disturbance of the stomach is avoided, as well as a liability to the production of abscesses, which not infrequently results when quinine is employed in this manner. After the vomiting has ceased, the good effects of the remedy are maintained by administering it by the mouth.

The following case will exemplify:

JULY 29th.—Infant, two months old. Has diarrhœa and vomiting; sick twenty-four hours; passages green and watery; vomits greenish fluid. Temp. 104⅞° in the rectum. Administered *sol. cinchonid. hydrobrom.* (*ten per cent.*) *m. viii.* hypodermically over the nates. Ordered, in conjunction with the above, the following:

℞. Acid. carbol. . . . gtt. ¼.
  Tinct. opii deodor. . . gtt. ¾.
  Mist. cretæ . . . . ℨi.
M. Sig. Dose, every two hours.

Cold baths given every three hours. Brandy and toast-water as a drink.

AUG. 5th.—Bowels improved. Vomiting checked. Temp 102⅖° in the rectum. Gave *cinchondiæ hydrobrom. gr. i. four times daily*, by the mouth. Afterward, the case did well.

DR. FRANCIS DELAFIELD finds that treatment of any kind is very unsatisfactory. Regarding the *diet*, if the child is nursing he usually allows it to continue, or in some cases he may change the nurse, or try hand feeding. Where, however, the child is bottle fed, he sometimes changes the food or directs that a nurse be obtained; but, as a rule, he finds that these cases do better with cream and water, or wine whey in small and very frequent doses. If vomiting is excessive so that all nourishment is rejected, inunctions of oil, olive or cod liver, are resorted to very effectually when other means fail.

For internal medication he knows of no remedy which, in his experience, has any very decided effect on the disease. *Opium* is given in small doses, and though it may not have much effect in checking the disease, yet it is of service in allaying the irritation of the child. In some cases he finds that *calomel in doses of gr. $\frac{1}{4}$-i. three or four times daily*, according to age, or mercury with chalk similarly given, seems to be very serviceable in checking the disease. In others, *ol. ricini* is given, not to act as a purge, but in small and frequent doses, a few drops in emulsion every three or four hours, and often seems to act very beneficially. Other remedies, such as the preparations of bismuth, *dilute hydrochloric acid*, or dilute sulphuric acid, are also used in

different instances. DR. DELAFIELD considers it
best, however, in all cases of cholera infantum, to
send the child to some other climate, and the earlier
the better. Nor in this respect, he advises, should
there be any hesitation, even through prostration is
so great that the child is apparently moribund.

After the acute symptoms of the disease have subsided, the following combination is often found remarkably beneficial:

R. Argenti nitrat. .   .   .   .   gr.i.
    Acid. nitric. dil.   .   .   .   m.viii.
    Tinct. opii deodor.   .   .   .   m.viii.
    Mucilag. acaciæ   .   .   .   ℥ ss.
    Syrup. simp.   .   .   .   .   ℥ ss.
    Aquæ cinnamomi   .   .   ℥ i.

M. Sig. A teaspoonful every three to six hours,
to a child of one year.

CONSTIPATION.

DR. A. JACOBI finds that young children with catarrh of the colon and follicular colitis, are very apt
to have a dilated colon ; and in these dilated artificial diverticula masses of fecal matter are retained,

the serum is absorbed and in this way the hard feces form, resulting after a time in paralysis of the colon and perhaps of the rectum and sphincter. In such cases of constipation, with paralysis of the sphincter ani and incontinence of feces, the discharges being hard and black, where no abnormal irritability of the intestine exists, he directs his treatment toward restoring the power of the sphincter. The general measures adopted by him, consist in the administration of nerve stimulants and tonics, such as strychnia. Locally, he resorts to the use of strychnia, applying after each act of defecation an ointment of the following;

R. Ext. nucis vomicæ . . . gr. xii.

Adipis . . . . . ℥ iii-iv.

M.

Or, he often uses the sulphate of strychnia hypodermically, thus:

℞. Strychniæ sulph. . . . gr.$\frac{1}{20}$.

Aquæ . . . . . q.s.

M. Sig. To be injected once daily.

He also employs the galvanic and faradic currents, using them alternately once a day. In addition to these means of treatment, he directs that *injections of cold water* be administed twice daily.

Regarding the *diet*, if there is also intestinal catarrh present, this is considered of the greatest importance. When milk is poorly digested, or not all, he directs that a little salt be added and the child be made to eat the milk, a spoonful at a time, instead of drinking it. Should even this be rejected, he finds that it may be retained by adding to the milk some glutinous substance, thus causing it to be acted upon very slowly by the gastric acids. In such cases, he usually gives very successfully a mixture of one-third to one-tenth of cow's milk, previously boiled, with oatmeal water, after boiling the latter alone for twenty minutes.

DR. JACOBI also calls attention to the fact that the size of the colon and its sigmoid flexure, as well as the number of its flexures, does not obtain its normal relation until from the fifth to the eighth year of life. In these flexures accumulations of feces often take place. Where this state of things exists, the secretion of the intestine is usually insufficient, causing a dryness of the feces resulting in constipation. Therefore constipation depending upon an abnormal length of the sigmoid flexure, and the presence of a number of flexures, will wear off, and is not amenable to treatment by purgatives, which, if used, may cause a paralysis of the sphincter; hence he advises that these should never be employed except in very urgent cases. When such a condition is suspected, he regulates the diet so that there may be an abundance of water in the food. For a food, he gives *oatmeal* in preference to any other. His main treatment, however, in these cases, consists in the regular daily use of *injections* into the bowels. These, he advises, are highly beneficial, but may have to be continued for many months; as the constipation, being anatomical, may not disappear until the cause is removed. He directs, therefore, that the intestine be washed out with simple warm water, day after day, while waiting for nature to restore the proper proportion of the intestinal canal. In obstinate cases he sometimes resorts to the use of the scoop.

DR. A. A. SMITH recommends the following very highly, in cases of constipation in children, as an exceedingly pleasant and efficacious laxative:

R. Ext. rhamni frangulæ fl. . · ℥ i.

Aquæ menth. pip. . . . ℥ i.

M. Sig. One to two teaspoonfuls, to a child from two to eight years of age.

WORMS.

For the expulsion of the *round worm*, DR. J. LEWIS SMITH uses the following very effectually:

℞.   Ext. spigeliæ fl.   .   .   .   .   ℥ i.

Ext. sennæ fl.   .   .   .   .   ℥ ss.

M.   Sig.   A teaspoonful, to a child of three to five years.

Or, in many cases he prefers to give the following, which he considers equally serviceable :

℞.   Ext. spigeliæ et sennæ fl.   .   .   ℥ i.

Santonini   .   .   .   .   .   gr. viii.

M.   Sig.   A teaspoonful, to a child of five years.

Moreover, he finds that these prescriptions are also an effectual means of destroying the *ascaris vermicularis.*

For the removal of the *tænia solium*, DR. SMITH often uses the ethereal extract of male fern with excellent results.   In many instances he administers :

℞.   Ext. filicis mar. æther.   .   .   ℨ i.

Sig. Dose, to a child of four years.

This amount he is in the habit of giving to these young children, and has never seen any bad effects follow.   In regard to the frequent failure in efficacy of this drug in tænia, he believes this is due to the fact that the dose prescribed was not large enough, and the patient had not been previously put upon the preparatory treatment which is so essential.

DR. A. JACOBI advises that in many cases the tænia found in very young children is the *tænia medio-canellata*, caused by feeding the child with raw meat as a remedy for diarrhœa; and finds that their expulsion is far more difficult than of the tænia solium in adults. To accomplish this, he administers *kameela powder*, ℨ *i. in the course of two hours.* Or, sometimes he prescribes kousso with good effect. Previous to this, however, he directs that the child be fed on oatmeal and milk, and purged three or four times daily. Then, after the administration of the kameela, a little milk is given, and if the worm does not appear in an hour, he orders a dose of castor oil. In his experience this plan is usually very successful.

DR. VAN GIESEN has found the aspidium marginale—the indigenous congener of the male fern—very serviceable in causing discharge of the worm, and generally in a short time. This he gives as follows:

℞. Ext. aspidii marginal. fl. . . ℨ ii.
Olei olivæ . . . . ℨ iv.
Æther. sulphuric. . . . ℨ ss.

M. Sig. Half a teaspoonful every fifteen minutes.

After three or four doses have been taken, he administers a purge of castor oil. This remedy, he thinks, in not a few instances, has undoubtedly done as well as the male fern usually does. In a great many cases, however, he also uses the *ethereal extract of filix mas in doses of m. xxx.* to a child of two years. This amount he gives very effectually, and without any bad results.

INCONTINENCE OF URINE.

At BELLEVUE HOSPITAL the combination of *ergot, belladonna, and iodide of iron* is employed in a large number of cases, and proves more successful for the incontinence óf urine in children than either of the drugs alone, or than any other combination which has been tried.

In many cases the following is often found to be of excellent service, affording speedy relief to the irritability of the bladder:

℞. Strychniæ . . . . gr. i.
    Pulv. cantharides . . . gr. ii.
    Morphiæ sulph. . . . gr. iss.
    Ferri pulv. . . . . ℈i.
M. Ft. pil. No. xl.

Sig. One, three times a day, to a child of ten years.

In conjunction with the above, a *cold shower bath* is usually given daily, and careful attention paid to the avoidance of irritant food and late suppers. The child should also be made to lie on the side or belly, taking care to drink nothing for a few hours preceding sleep, and to empty the bladder thoroughly before going to bed.

INFANTILE LEUCORRHŒA.

DR. T. GAILLARD THOMAS considers it of primary importance to ascertain if worms, usually the ascaris vermicularis, are present; and if so, or should they be suspected, he uses an injection of warm salt water. He then endeavors to build up the general nutrition of the child, and place it in the best possible condition, by appropriate food, iron, vegetable tonics, and the administration of the hypophosphites. He places more dependence, however, upon *a good nourishing diet* than upon medicines. After the expulsion of the worms, should the irritation and discharge continue, or even if no worms have been present, he then resorts to local measures. Placing the child upon the back in order that the canal may be perfectly cleansed, he washes out the vagina thoroughly, by means of a syringe provided with a small nozzle which has been previously well oiled. In many instances, he finds that the mere removal of the accumulated secretion, which is a constant source of irritation, is all that is required. Where, however, the trouble has lasted for some time, this is not always sufficient. In such cases he uses the old-fashioned black wash, as follows:

℞. Hydrarg. chlor. mit.   .   :   ℥ ss–i.
    Aquæ calcis .   .   .   .   . Oi.
M. Sig.   To be applied twice daily.

This he believes to be one of the best applications that can be employed. Before using it, an injection of simple warm water is made.

Under this treatment, DR. THOMAS has never failed to cure infantile leucorrhœa in a short time; therefore he advises against any necessity of resorting to *astringents* and *silver nitrate*, which may do harm. He also directs that the mother or nurse be instructed how to properly introduce the nozzle of the syringe, otherwise it is seldom carried far enough up into the vagina, while it should reach the upper part, and improvement will fail to follow its use.

DR. J. B. HUNTER generally finds constitutional treatment and bathing of the external genitals sufficient. As a final resort, he recommends *injections*. In this respect, however, he objects to cold water injections as injurious, having on many occasions seen bad results from their use.

DR. V. P. GIBNEY has frequently met with cases of sero-purulent vaginal discharge, in children of two to six years of age. In some instances he places the child upon the following, with much advantage:

℞. Calcii sulphidi . . . gr. $\frac{1}{10}$.

Sig. Dose, every three hours.

If no benefit obtains in a few days, he orders warm and slightly *carbolized injections*. When indicated, he also prescribes iron and cod liver oil. Sometimes he finds this plan is very serviceable, while at others no benefit at all is apparently derived.

DR. BEVERLY ROBINSON is not inclined to ascribe so much importance to the condition of diathesis, in cases of leucorrhœa in children. Local treatment is considered by him as generally much more effectual. Moreover, in the treatment of vaginitis by injections, he thinks that *cold injections* probably have an advantage over warm ones.

At the NEW YORK HOSPITAL this condition is found to be not infrequently associated with a strumous diathesis; or, occasionally the urine exhibits the presence of a large amount of uric acid. In some of these cases local measures are adopted, although the propriety of this treatment is considered as doubtful.

PART IV.

DISEASES OF THE BRAIN AND NERVOUS SYSTEM.

CONVULSIONS.

DR. J. LEWIS SMITH considers the treatment of infantile convulsions one of the greatest importance, as these cases are so frequently met with without being able first to ascertain the cause. Ordinarily, regarding the seat of the disturbance as being in the intestine—either overloading or indigestible food— he first causes evacuation of the bowels. For the convulsion itself, when occurring in infants and dependent upon pain produced by the irruption of the teeth, he rarely or never uses the *gum lancet*. His reliance is almost wholly upon the bromides, preferring the potassium salt in large and frequent doses, thus:

℞. Potass. bromidi . . . gr. ii–vi.

Aquæ . . . . . · q. s.

M. Sig. Dose, every ten, fifteen, or twenty minutes, to a child between two and twelve months of
age.

This he repeats as above, according to the severity
of the case and to the control of the irritation, and
he has never seen any bad results follow from this
course.

In the convulsions of children, his plan is to give
chloral hydrate by rectal injection, combined with
the internal use of the bromide, and if the spasms
are not soon controlled he administers chloral hypodermically. He usually prescribes the following:

R. Chloral. hydrat. . . . . 3 i.

Aquæ . . . . . . ℥ iss.

M. Sig. 3 i. (gr. v.) every five or ten minutes, as
an enema.

By this treatment, if retained, he finds that the
beneficial effect of the drug is often surprisingly
shown, and its action readily and speedily exhibited.
The convulsive movements are controlled in the
shortest possible time, generally within twenty
minutes, and in many cases in even a shorter period.
He then continues the treatment with the bromide,
if necessary. As to the administration of anæsthetics, that is to say of *chloroform*, DR. SMITH never
resorts to this agent. He altogether prefers *chloral*

hydrate, thus not only avoiding the danger of paralyzing the respiratory center, but also of being com-
pelled to remain three or four hours with the patient.
Concerning the *hot bath*, he often finds it an exceedingly useful adjunct.   In cases, so frequently met
with, where convulsions are threatened, with sudden
twitching, starting, etc., and especially if the hand is
affected or the thumb turned in, he has repeatedly
found that by placing the child in a tepid bath of
100°, a soothing effect is obtained and sleep often
induced after its use.   The hot bath, however, may
in some instances, be too stimulating, but he finds
no objection in applying it to the extremities, as in
the majority of cases active or passive cerebral congestion is present, which may be thus relieved.

In cases of convulsions, reflex in character and
dependent upon intestinal irritation, DR. A. JACOBI
gives a purgative dose of calomel, following this in a
few hours afterward by:

    ℞.   Chloral. hydrat.   .   .   gr. iv.

        Potass. bromidi   .   .   .   gr. viii.

        Aquæ   .   .   .   .   .

        Syrupi   .   .   .   .   .   āā q.s.

    M.   Sig. Dose, to a child of two years.

This, he finds, together with attention to diet and
proper hygiene, is usually sufficient.   During the
convulsive spasms he also cautions that the child

must not be touched, since such disturbance may increase their severity, or give rise to others when the patient is recovering.

In the nervous disturbances and excitements of children, DR. A. A. SMITH uses the bromides very beneficially. He advises, however, that the administration of bromide to young children is often difficult, on account of the disagreeable taste attending its use. This he very effectually obviates by giving the drug in small and frequently repeated doses; usually selecting the sodium salt for this purpose, as being less nauseating, and administering it as follows :

℞. Sodii bromidi . . . gr. xxx–lx.

Aquæ . . . . . . ℥ iv.

M. Sig. A teaspoonful every ten or fifteen minutes.

When given in this manner, he finds that the remedy often renders most valuable service in the nervous affections arising from dentition and other causes, and in relieving the fever which, in children, usually attends any slight degree of excitement.

Also, in those cases of highly nervous, fretful, and excitable children, so frequently met with, where there is inability to sleep at night, the indication being to administer a sedative, he recommends the following, which has been used by him with much satisfaction :

℞. Tinct. chamomillæ   .   .   m. viii

Aquæ .   .   .   .   .   .   ℥ i.

M. Sig. A teaspoonful every fifteen or twenty minutes.

This, he advises, is also a tonic as well as sedative, and in these cases acts far better than chloral hydrate which is often liable to affect the digestion.

In the convulsions of childhood, during an attack, whether due to organic disease or to functional disturbance, DR. SMITH directs that the convulsion be arrested at once and another prevented by the administration of anæsthetics, preferably *chloroform*. If dependent upon pain produced by other than causes such as the pricking of a pin, tight abdominal bandage which can be easily removed, or an overloaded stomach which should be emptied by an emetic, he considers *opium* the most valuable remedy, employing it in all children over four months old.   Where the convulsion depends upon the pain of teething, he first controls this by opium, and then uses the gum lancet.   Regarding *lancing the gums*, he believes that if they are swollen and hot, they should be lanced; or, if it is time for the tooth to appear, he also resorts to scarification, as he finds that the irritation is often due to deepseated pressure which is not manifest upon the surface.   In such cases, according to his experience, lancing the gums is frequently followed by marked relief.

The same principle is embodied in his treatment of convulsions due to worms, or other foreign bodies in the intestinal canal. Opium is first administered and then a *cathartic* given; or sometimes he combines the opiate and cathartic with advantage. Where the source of irritation is in the rectum, or near it, after the influence of opium is sufficiently secured to control the convulsion, an *enema* is given. In convulsions due to malarial poison, having arrested the paroxysm by opium, he puts the child fully under the influence of *quinine* to prevent its recurrence. In this latter class of cases, he advises that the tolerance of opium is sometimes very great.

Therefore, as in most instances, opium is indicated, he directs the mother to give the baby, if over four months old, paregoric, with explicit instructions regarding the size of the dose, to be repeated every half hour until the convulsion ceases, or a physician arrives. Where the child is under four months of age, he prescribes the following with best effect :

℞. Potass. bromidi . . . .
    Chloral. hydrat. . . .
    Sodii bicarb. . . . āā gr. xvi.
    Aquæ ferv. . . . . ℥ ii.

Sacchari   .   .   .   .   .   q.s.

M. Sig. A teaspoonful every hour.

Many times, however, in infants from six weeks to
four months old, he gives double the quantity every
hour, or every two hours, according to the frequency
and the violence of the convulsions.

Regarding the use of the *hot bath*, DR. SMITH
finds that one or more spasms almost invariably
occur while the child is in the bath, owing to the
excitement, etc., in administering it ; hence, as it is
necessary to the treatment to avoid all agitation, he
objects to its employment. He moreover insists
that the child be not restrained while in a convulsion, but that it be placed in bed in a room which is
kept perfectly quiet, partially darkened, and supplied
with plenty of fresh air. The opening and shutting
of doors are also to be avoided, and only one person
permitted to remain in the room at a time. These
minute particulars he considers essentially of great
importance. He further advises that all over-active
treatment is not only uncalled for, but may be positively dangerous.

In those instances where the prolonged use of
opium is undesirable, he believes the *bromides* to be
of marked value; their sedative effect must, however, be maintained. He finds their action very
useful in averting threatened attacks; also particu-

larly favorable in the cases of convulsions in young children, apparently due to the intense itching which especially follows scarlet fever and measles. Where, however, the symptoms are aggravated by the bromides, he uses *chloral* with good effect. In the convulsions of hooping cough, he gives the bromides and chloral in combination. If associated with disease where there is a tendency to heart failure, or with the exhaustion of severe diarrhœa, he resorts to *stimulants*, preferably musk and camphor. In convulsions depending upon high temperature, he administers veratrum viride, as follows:

℞. Tinct. verat. viridis . . gtt. ii.

Sig. Dose, every hour or two, to a child of six to eighteen months.

Or, he sometimes gives it in combination with opium, thus preventing its tendency to cause vomiting. Where this remedy fails, however, he orders a cold bath, as the most efficient means for this object. For the convulsions occurring at the onset of an acute pulmonary affection, he finds the administration of *calomel in sedative doses of gr. v.*, to a child of one to three years, often very serviceable. This he also combines with the use of veratrum viride and the cold bath, if necessary.

Regarding the subsequent treatment, DR. SMITH advises that the child be kept in the best possible condition, at the same time directing his treatment to the removal of the causes, as well as of the pre-disposing causes; *e. g.*, the rickety diathesis, syphilis, rheumatism, disturbance of digestion, nervous excite-ment, etc.

In the spasms occurring in young girls at about the age of puberty, and evidently due to a diffused pelvic irritation, no dyspepsia or local disease being present, DR. E. C. SEGUIN finds medicinal treatment somewhat uncertain. He sometimes derives benefit from giving the bromide of potassium in moderate doses, thus:

    ℞. Potass. bromidi   .   .   .   gr. xx.

    Aquæ   .   .   .   .   .   . q. s.

    M. Sig. Dose, every night.

In addition to this he applies a blister to the back of the neck, with much benefit. In other cases, however, this treatment entirely fails of relief. Again, he uses *ergot*, giving it in view of the theo-retical cause of fullness of the ovarian plexuses, and has at times found it very serviceable.

For *local action* in controlling infantile convul-sions, the following is very highly recommended by some, and has been found to almost invariably arrest the paroxysm:

    ℞. Olei succini rectif.   .   .   .

    Tinct. opii   .   .   .   .   . āā ℥ ss.

Olei olivæ . . . . .
Spts. vini gall. . . . . āā ʒ ii.
M. Ft. lotio.
Sig. Rub along the spine.

Before the application is made, it is directed that especial care should be taken to see that the skin is well washed with warm water and soap, in order to promote rapid absorption.

CHOREA.

Dr. J. Lewis Smith finds that under good hygienic conditions this disease usually terminates in a few months. This termination he endeavors to hasten by suitable treatment. If the child attends school, especially public school, the disease is apt to be aggravated by the severe discipline attached to these institutions, therefore he advises remaining at home for a season. For internal treatment, where no organic lesion is present to account for the chorea, and considering the case purely as a neurosis, he usually derives much benefit from placing the child on the use of arsenic, giving:

℞. Liq. potass. arsenit. . . . gtt. v.

Aquæ .   .   .   .   .   .   q.s.

M. Sig. Dose, three times daily, to a child of eight to ten years.

This is gradually increased to gtt. viii–x. three times daily.

Or, in other cases, he gives Fowler's solution, m. ii. after each meal, together with the following :

℞. Ferri et potass. tart.   .   .   . ʒ i.

Tinct. cinch. co.   .   .   .   . ℥ iv.

M. Dose : A teaspoonful three times daily.

With these measures, he insists that the *hygiene* of the patient should be strictly attended to, the bowels kept open, plain nourishing food given, and pure air insured. By careful attention to these means improvement rapidly follows.

In cases of chorea associated with organic heart disease, with anæmia, etc., for general treatment, where there is little or no fever, DR. A. JACOBI administers a course of *iron*. For this purpose he prefers the subcarbonate or the *pyrophosphate, in doses amounting to gr. xx–xxv. daily*. The tincture of the chloride and the muriated tincture are avoided, from being vascular irritants. For the chorea, he places most reliance upon arsenic, using either the sodium or potassium salts, as follows :

℞. Sol. Fowlerii   .   .   .   . gtt. iii.

Aquæ . . . . . . q. s.

M. Sig. Dose, three times daily, after meals.

This amount he gradually increases to gtt. vi–viii. In all cases, he believes it to be of the greatest importance that these children should rest well at night; hence in severe forms of the disease, when the muscular twitchings persist disturbing and interrupting the sleep, he endeavors to control this condition by the administration of chloral and the bromides, thus:

℞. Chloral. hydrat. . . . gr. xv.

Potass. bromidi . . gr. xv–xxiv.

Sig. Dose, at bedtime, to a child of eight to ten years.

If necessary, to accomplish its object, he increases the chloral to gr. xx–xxv.

DR. W. A. HAMMOND finds the following often of excellent service:

℞. Zinci bromidi . . . . ℨ i.

Syrupi simp. . . . . ℥ i.

M. Sig. Ten drops, three times daily.

In using this, he advises that the dose must be increased as rapidly as the stomach can bear it; then, with the disappearance of the chronic symptoms, it is to be gradually lessened. He also prescribes the following in many instances, and considers it highly efficacious:

℞. Strychniæ sulph. . . . gr. ii.

Aquæ . . . . . ℥ i.

M. Sig. Five drops, three times daily, to a child of ten to fifteen years.

DR. E. C. SEGUIN recommends *hyoscyamus* with much satisfaction. Regarding the administration, he finds that it can be given in small doses, hypodermically, with ease, and distinct effects obtained from gr. $\frac{1}{100}$. The following formula is generally used by him:

℞. Hyoscyamiæ (Merck's crystallized) gr. i.

Glycerinæ . . . . . .

Aquæ destill. . . . . āā m. c.

Acidi carbol. pur. . . . gtt. i.

M. Filtra.

Sig. m. i. contains gr. $\frac{1}{200}$.

Of this he considers m. ii. as a moderate dose, and m. iv. a full dose. For administering by the mouth, he uses tablets containing gr. $\frac{1}{50}$ very conveniently. He has also found the *fluid extract of conium* very beneficial; giving it in large doses for the purpose of obtaining the paralyzing effects of the drug.

As a rule, however, arsenic is most uniformly successful. In cases of post choreic paralysis he places great reliance on *Fowler's solution in doses of gtt. v.* at first, and gradually increasing. If the chorea is violent, chloral is also given. By this means, together with extra feeding, rest, and rubbing, DR. SEGUIN finds that these cases of even very grave paralysis will get well within two or three weeks. On the other hand, he cautions that if large doses of arsenic are given, the patient may linger for months. In his experience the *cold shower, chalybeates* and *cold water* do not act as quickly in chorea as the thorough arsenical treatment.

In choreic children, DR. W. H. THOMSON considers zinc and arsenic most serviceable; of these he prefers the former, giving *zinci sulphat. gr. i. three times daily*, and increasing till nausea is produced. From this treatment he has had marked improvement within a week. For the abnormal condition of the nervous system, with faulty nutrition, impaired memory, etc., he prescribes the following with benefit:

R. Ferri lactat. . . . .
 Quiniæ sulph. . . . āā gr. xx.
 Pulv. carbon. . . . . gr. xl.
M. Ft. pulv. No. x.
 Sig. One, three times daily.

This he directs to be taken in cod liver oil. He further advises, that even though great improvement in the chorea takes place, yet the general treatment is to be continued for several months.

DR. E. G. JANEWAY, as a rule, derives most success

from *strychnia*. Where there is also marked anæmia present, he places the child on the use of iron, in small doses, and continues it for some time.

At the NEW YORK HOSPITAL the following arsenious acid tonic is continually employed, and is considered to be a most efficient combination:

℞. Acid. arseniosi . . . . gr. ¼.
Ferri et quin. cit. . . gr. lxxx.
Tinct. cinchon. co. . . . ℥ ii.
M.

EPILEPSY.

To ward off the immediate attack, DR. W. A. HAMMOND directs that when the child feels an aura, *gtt. iii–iv. of amyl nitrite* be put upon a handkerchief and inhaled. This he has often found very efficacious, and has frequently succeeded not only in aborting a paroxysm, but in curing several cases of epilepsy by this method alone, and without the administration of any internal remedy. In many instances, however, he finds that even when aborted by amyl nitrite, the paroxysm is very apt to recur. In using it, he advises that the drug should be put to the mouth so that it may be inhaled thoroughly, and produce a sense of fullness in the head, a tingling of the surface of the body, and redness of the face. In certain cases, also, where the bromides have failed, he has derived much benefit from using nitrite of amyl internally, as follows:

R. Amyli nitriti . . . . gtt. i.

Sacchari . . . . . q. s.

M. Sig. Dose, three times daily, increased gradually to gtt. iii–iv.

He has also obtained excellent results from *nitroglycerine* (or glonoin) in this malady, and finds its effects more permanent than amyl nitrite, although it takes a longer time to act. In the beginning of an epileptic attack, when used for the same purpose as amyl nitrite, he is able to effectually prevent the spasm of the arterioles by this means. Moreover, he has employed nitro-glycerine quite successfully for the permanent treatment of the disease. In the status epilepticus, and in cases of long standing, where the bromides have signally failed, he often finds this remedy very effective. Or, in some instances, he gives it in conjunction with the bromides. Regarding administration, the strength usually employed by him is a one per cent. solution, of which he prescribes:

R. Nitro-glycerin. . . . gtt. i.

Alcohol .   .   .   .   .   gtt. c.

M. Sig. One drop three times daily, on a lump of sugar.

This amount, he advises, is borne very well by children. After a time he gradually increases the administration to gtt. v. as required. Under this treatment, he finds that some cases improve remarkably, and not infrequently the attacks appear to be permanently stopped by this agent. Or, he frequently administers this drug in pill form, of which he prefers Metcalf's preparation.

As a rule, however, and especially where the bromides have not previously been given, DR. HAMMOND considers it best to begin with their administration in some form. Regarding the old remedies, such as the salts of *zinc*, etc., though sometimes beneficial, yet he finds them more so in conjunction with the bromides than when used alone. On the other hand, he advises that the *bromides* will almost invariably reduce the frequency of the paroxysms, and if the case is not an old one, they may effect a permanent cure. For administration he prefers the sodium salt, as it is much more pleasant to the taste, and if desired may be taken with the food. In his experience, the following is considered to be one of the best plans of treatment, and from which he has obtained uniformly good results; he insists, however, that it must be rigidly carried out in all its details. In prescribing this treatment he usually begins with :

℞. Sodii bromidi . . . . ℥ iv.

    Aquæ . . . . . Oi.

M. Sig. A teaspoonful (gr.xv.) three times daily, to a child of eight to twelve years.

He further directs that each dose is to taken largely diluted with water, as by this means the effects of the drug are greatly increased. The efficacy of the above solution, he also advises, will be much enhanced by the addition of *potassium iodide*, ℥ ss.

Regarding the results of this method, he finds that, as a rule, it will take several days for the drug to produce its effects, as it acts very slowly. He then increases each dose by one-fourth every three months for a year, and continues it at that amount for another year ; after which he commences to reduce in the same manner, so that during the fourth year the patient takes gr.xv. of the bromide three times daily. In this respect, however, he cautions that if the quantity is not increased during the first year, the attacks will probably recur, thus adding to the difficulty of arresting them. Formerly, DR. HAMMOND used to consider that two, or possibly three, years was long enough to continue giving the bromides ; but of late he believes that a relapse is thus encouraged, hence

II

he advises that it is safer to continue the treatment for four years. If, however, after a time, the bromides do not produce so good results as are desired, he stops their administration for a month or longer, until the system has become entirely free from the drug, and then begins the treatment anew; in the meantime giving *cod liver oil and tonics*, to quiet the nervous irritability, etc. Moreover, in all cases, he insists that unless a certain degree of bromism is produced, the disease cannot be cured. Any weakness short of inability to stand up, and an acne eruption on the face and chest, are, in his opinion, not contra-indications to a continuance of the treatment.

To prevent the eruption of acne, he sometimes combines *Fowler's solution* with the bromide ; but, as a rule, when injurious effects are produced, he finds that the best plan is to diminish the dose. The combination of *iron* with bromide in epilepsy, he believes to be harmful. As to *quinine*, in his experience the use of this drug in combination lessens the effect of the bromides upon the patient.

In certain cases he finds that indolent ulcers are caused by the long use of the bromides ; these he usually easily cures by galvanism. Or, in a great many instances, *pure bromine* is employed by him instead of the salts, by which means the acne and the ulcerative manifestations of the bromide are, he believes, avoided. His formula for this administration is:

℞. Brominii . . . . . ℨi.

Aquæ . . . . . ℨviii.

M. Sig. A teaspoonful, well diluted with water.

Another useful measure which DR. HAMMOND resorts to very satisfactorily, in the treatment of epilepsy, is *counter-irritation*, applied to the back of the neck by means of a platinum disc, or other instrument, heated to a white heat. In doing this, he advises that it is necessary only just to touch the skin, and then remove the cautery immediately; the pain thus produced is so slight that the patient scarcely feels it. By this procedure, he has found the number of paroxysms to be reduced after a single application.

DR. J. LEWIS SMITH urges the great importance of making a correct diagnosis early. If this is done, he usually controls its manifestations by the free use of the bromides, giving the following:

℞. Potass. bromid. . . . ℨss.

Aquæ pur. . . . . ℨv.

M. Sig. A teaspoonful three times daily, in cold water.

This amount is gradually increased as indicated.

DR. E. C. SEGUIN believes that attacks of the

petit mal, or epileptic vertigo, though seemingly an insignificant symptom, is nevertheless far more serious than the grand mal, or fits, and does not always receive the attention which its existence demands.   He therefore considers that the importance of its early recognition cannot be magnified, and its existence should be met with prompt and careful treatment, at the earliest possible moment.   The *bromides* are prescribed by him, together with atten_ tion to the digestive apparatus, etc. ; although he finds that these cases, and especially the flash-like form, are exceedingly rebellious to treatment, and many children will continue to have several turns a day, despite the administration of as much bromide as their systems will bear.   In several instances he has found it necessary to produce severe bromism, in order to barely control these minor forms of epilepsy ; and the least reduction of the medicine to a safer dose, has been followed by a return of the symptoms. In all forms of the disease, however, both in grand and in petit mal, he resorts to early treatment, if possible after the first or second attack, as by this means he is often able to eliminate the epileptic habit, and thus greatly increase the chances of cure.

Regarding the convulsions, in children under three years of age, he advises,the first convulsions may probably be eclamptic, but the causes of the attack should receive the most careful judgment. In these cases, that is to say, which do not present an evident eclamptic condition, he finds it well to give a moderate amount of potassium bromide with regularity, for several months after a convulsion. After the third year he believes the attacks to be, as a rule, epileptic, not forgetting the possibility of uræmia and syphilis. In such children, he at once institutes the bromide treatment, and continues it for many months. This plan, he advises, does not interfere with the treatment by appropriate remedies and by hygiene, of gastric or intestinal indigestion, of worms, or of other abnormal conditions. Moreover, in his experience, even the most severe fits can nearly always be controlled by a proper dosing of the bromides, and may also be suspended for a period of months and years.

In giving the bromides, DR. SEGUIN always endeavors to ascertain the minimum dose which will control the seizure; this amount is then increased as demanded. For administration, to commence with he generally prescribes:

℞. Potassii bromidi . . . ℥ i.
Aquæ . . . . . ℥ vii.
M. Sig. ℨ iii–vi. daily.

He also directs that each dose be diluted with *vichy water*, as in this way the taste is somewhat disguised; or, when this cannot be obtained, he finds that a pinch of soda in the water answers very well. Where there is much depression from the use of the bromides, he gives *strychnia* with considerable efficacy. In some cases he has also found *iron* to act very beneficially. When indicated, he combines chloral with the bromide, and often with remarkable effect; thus:

℞. Potassii bromidi   .   .   .   ℥ i.
Chloral. hydrat.   .   .   .   ℥ ss.
Aquæ   .   .   .   .   .   ℥ vii.

M. Sig. ʒ iv–vi. daily, in water.

Again, in numerous instances he prefers to give the following:

℞. Ammon. bromidi   .   .   .   ℥ ss.
Potassii bromidi   .   .   .   ℥ i.
Aquæ   .   .   .   .   .   ℥ vii.

M. Sig. ʒ iii-iv. daily, increasing the amount as required.

Furthermore, besides the use of the bromides, he finds that a variety of treatment is frequently required, according to the pathological condition of each individual case.

INFANTILE PARALYSIS.

In those cases of paralysis which come on without any apparent premonitory symptoms, DR. A. JACOBI relies a great deal upon the *galvanic and faradic current*, and upon *strychnia* both subcutaneously and internally, preferably the former, from which, also, he obtains better results by applying the injections directly over the affected part. For the same reason, in paralysis of the sphincter ani due to muscular relaxation, he uses the alcoholic extract of nux vomica, as an ointment, with the most beneficial effect. Of this, the following strength is usually employed by him:

℞. Ext. nucis vomicæ . . gr. xii.

Unguent. . . . . ℨ iii–iv.

M.

In so-called dental, or spinal, paralysis, to subdue the local inflammation, he advises the application of *ice* to the surface for hours or even days; not forgetting, however, that infants and young children bear the protracted application of cold badly. For this reason, he usually keeps the ice in contact with the surface for an hour or two, and then, after re moving it for the same period of time, repeats the application; continuing in this manner as long as necessary. If constipation is present he relieves it by giving *calomel in doses of gr. v–x.* according to age; or sometimes a saline is used. In these cases he always uses ergot in large doses; as a rule, giving:

℞. Ext. ergot. fl. . . . ℨ i.

Sig. This amount in twenty-four hours, to a child of one year.

This administration he continues for three or even six weeks. Or, at times, larger doses are prescribed by him; in some instances as much as ℨ ii. In other instances, he prefers to use *ergotin by hypodermic injection.* Where the heart is feeble and requires stimulating, digitalis is employed. After the fever and pain and all active symptoms have disappeared, he supplements the ergot by the addition of *iodide of potassium to the amount of* ℨ *i.* in twenty-four hours. As regards *inunction*, he considers the ointment of potassium iodide of little or no service. In his experience glycerine and potassium iodide act better, and he has often observed, after such an application, the urine impregnated with iodine within twenty-four hours. He has also found iodoform ointment exceedingly beneficial, or better, the following:

℞. Iodoformi . . . pars i.

Collodion   .   .   .   part. xii–xx.

M. Sig. To be applied three or four times daily.
Besides, this being non-irritating, he finds that it can be continued for weeks with no ill effects.

Regarding pseudo-hypertrophic paralysis, DR. JACOBI has succeeded in curing a case by means of the *hydrarg. corros. chlor. in doses amounting to gr.* $\frac{1}{40}-\frac{1}{50}$ in twenty-four hours, and continued for a period of eighteen months.   Under this treatment, nutrition was not impaired in the least from the effects of the drug, but on the contrary the child ate and slept well, became rosy and healthful, and the atrophy and hypertrophy finally both gave way.

DR. F. DELAFIELD'S indications for treatment are to support the strength and nutrition, and render the patient as comfortable as possible. Any derangement of digestion is carefully attended to, and the bowels kept freely open.   While the disease is progressing, he considers it a good plan to apply *dry cups* along the spine ; but after this stage has passed, and the disease has reached a systematic development, he obtains marked benefit from using the *faradic current*, and in thoroughly *rubbing* the muscles at stated periods.

DR. W. A. HAMMOND considers ergot to be of excellent service, in cases of infantile spinal paralysis before the stage of atrophy has set in. He generally prescribes :

℞. Ext. ergotæ fl.   .   .   .   ℥ i.

Sig. Ten drops three times daily, to a child of six months.

In conjunction with this, local measures are also employed. After the period of atrophy is reached, he then administers the following :

℞. Strychniæ sulph.   .   .   .   gr. i.

Ferri pyrophosph.   .   .   .   ℥ ss.

Acidi phosphorici dil.   .   ℥ ss.

Syrupi zingiber.   .   .   .   ℥ iiiss.

M. Sig. m. xx. three times daily, to a child of two years.

From this he has often derived marvelous effect.

In infantile hemiplegic paralysis, DR. E. C. SEGUIN varies his treatment according to the pathological diagnosis. *e. g.* The treatment of a case depending upon convulsions from injury, will differ materially from one in which tubercular meningitis is suspected; or one in which some gastric irritation is thought to exist. In the first stage of hemiplegia in children, he finds that it is best to avoid doing too much. The only exception to this, he advises, is in cases where, at an older period than mere babyhood, convulsions occur followed by hemiplegia of a rather slowly developed kind, and continued convulsions in which the teeth are bad, etc.   In these cases his general rule not to do too much is reserved.   Mercury, potassium iodide, and cod liver oil will, he finds, do a wonderful amount of good.

Regarding treatment of the sequelæ, nothing can be done for the pathological degenerative changes in the brain.   Various clinical sequelæ are however met according to the indication.   The athetoid and choreic conditions he endeavors to correct by gymnastics, massage, and electricity.   Epilepsy, which is often present, is treated by the *bromide of potassium ;* but he advises that, in these cases, the drug is not required in large doses as in common epilepsy.   Or, in many children who exhibit the remains of hemiplegia and epilepsy at the same time, DR. SEGUIN has obtained excellent results from the combination of the iodide and bromide of potassium.   In such cases, he generally gives :

℞. Potass. iodidi . . . . gr. v.

Potass. bromidi . . . gr. x.

Aquæ . . . . . q.s.

M. Sig. Dose, twice daily, to a child of four or five years.

At the same time, he directs that the general condition, the hygiene, etc., should receive the most careful attention.   Where imbecility exists, this is to be corrected by judicious training ; and in the

case of the poor, he advises that the parents be instructed to send their children to the schools for
idiots.   For whatever deformity that may be present
he aims to remedy this by proper treatment, partly
by massage, by electrical treatment, exercise, and
also by orthopædic apparatus.   In regard to *electrical
treatment*, he advises that this must be applied to
the weakest muscles.   If the whole arm is faradized
the child is made worse.   Where, however, the
faradization is limited to the extensors and interossei,
he often obtains a marked change in the child's hand
and arm in the course of a few weeks.   Moreover,
the current should be strictly localized ; the elec-
trodes are to be placed close together, so as to produce contraction in all the weakest muscles, the
flexors requiring no treatment.   *Massage* and passive
movements are also considered very important.   In
some cases, the maintenance of the normal attitude,
or even over-extension in the clenched instances by
a little apparatus, will, he finds, do a great deal of
good.   The child is also forced to use the hands and
legs, at intervals ; the well hand being tied up for a
few hours a day, according to the temperament of
the child.   Finally, what in the education of idiots
is called *physiological training of muscles*, is considered of the greatest importance, and should be
attempted in all cases ;   *e. g.*, training the hand for

delicate movements, by a graduated series of little muscular exercises.

Therefore, under this plan of treatment, carefully and persistently carried out, if the circumstances are favorable, DR. SEGUIN finds that a great deal can be done for these children.

At BELLEVUE HOSPITAL the following solution of ergot is employed for hypodermic use :

R. Ext. ergotæ e fl. (Squibb) . gr. lx.

Aquæ q.s. ad . . . m.ccc.

M. Sig. m.i. contains gr. i.

A similar preparation is also used at the NEW YORK HOSPITAL.

At CHARITY HOSPITAL the following is administered :

· R. Ergotini . . . . gr. xxxvi.

Glycerinæ . . .

Aquæ . . . . āā m. cviii.

M.

ACUTE MENINGITIS.

To subdue the local inflammation, DR. F. DELA-FIELD advises that the employment of blood-letting and cold will depend upon the age and constitution of the patient, and the period at which the case is seen. To a strong, full-blooded child in the early stage of the disease, he administers a brisk purge, and then proceeds to draw blood by *wet cups* or *leeches*, applied over the temples and back of the neck. This procedure, he advises, must be resorted to at the commencement and during the first forty-eight hours of the disease, when he finds it of excellent service; after that period has passed, however, no benefit is derived from its use. He also employs *cold* in the shape of ice-bags, applied by cutting the hair and covering as much of the surface of the cranium as possible; and by placing an ice-bag beneath the neck and adjacent part of the head. This application of cold he finds most serviceable during the first two or three days, but continues its use with much benefit throughout the first week. *Opium* is also administered in some form, to quiet the restlessness and delirium.

If the case is not seen until five or six days have elapsed, he considers opium, cold, bleeding, and all depleting measures contraindicated, and should be avoided. At this stage, he administers *iodide of potassium* in moderate doses, and supports the strength of the patient with quinine, iron, etc.

When the disease occurs as a complication, or in asthenic children, he objects to blood-letting, but applies cold to the head and administers opium to relieve the pain ; or, should opium be contraindicated, he gives potassium bromide or chloral, or both. A purge is also given unless the bowels are already relaxed from the concomitant disease.

In cases of basilar meningitis, DR. E. C. SEGUIN derives excellent results from iodide of potassium, its administration often causing amelioration of the symptoms, with rapid general improvement and, in many instances, final recovery. Moreover, in these cases he believes syphilis may often be entirely excluded. In conjunction with this, he also resorts to counter-irritation by means of blisters, applied to the back of the neck. Regarding the use of potassium iodide, to accomplish its object, he finds that large doses may be required and can be given with apparent impunity. Even in young children of eight to ten years, he has, in some cases, found it necessary to give the following :

℞. Potassii iodidi . . gr. xxx–cl.

Aquæ . . . . . q.s.

M. Sig. Dose, three times daily.

DR. W. A. HAMMOND has also obtained like benefit from this remedy. In many instances, however, he advises that enormous doses may be necessary before success follows. Therefore, to facilitate this administration, when prescribing the drug he directs that a great deal of water be taken with each dose, sometimes a pint, as in this way he finds that these large doses may be given without any inconvenience and iodism prevented.

ACUTE HYDROCEPHALUS.

(TUBERCULAR MENINGITIS.)

———

DR. F. DELAFIELD directs his treatment chiefly to the symptoms. If the urine is scanty or suppressed, he orders a *hot bath* for the space of five minutes, after which the child is immediately wrapped in a blanket, without waiting to dry the body, and thus kept in a sweat for one, two, or more hours. Diuretics he finds of little use, although he sometimes gives the iodide or acetate of potassium. For the irritability and restlessness, he administers the bromides with chloral; thus:

℞. Potassii bromidi   .   .   .   gr. i–v.
    Chloral. hydrat.   .   .   .   gr. i–iii.
    Aquæ   .   .   .   .   .   .   q. s.
M. Sig. Dose, every two or three hours.
In addition to this, to act on the inflammation, he

uses *potassium iodide in doses of gr. ii–v.* according to the severity of the case. Locally, *counter-irritation* is resorted to by means of blisters applied to the neck. He also applies cold to the scalp, abstracts blood from the temples, and produces free catharsis. For internal medication DR. DELAFIELD further administers *ext. ergotæ fl. in large doses.* Where convulsions are very marked, chloroform may be resorted to chiefly for the mother's sake.

In the earlier stages, before effusion has taken place, the two remedies upon which DR. J. LEWIS SMITH places most reliance are ergot and potassium iodide. If nervous symptoms are developed he gives bromide of potassium in addition. He usually prescribes:

℞. Potassii iodidi . . . gr. xvi.

Aquæ . . . . . . ℥ i.

M. Sig. A teaspoonful every hour, to a child of one year.

With this he gives *fluid extract of ergot (Squibb's) gtt. vi–viii. every three hours.* As regards post aural vesication, although it does no harm, yet he believes that very little good results from resorting to it.

SPURIOUS HYDROCEPHALUS.

See CHOLERA INFANTUM.

CHRONIC HYDROCEPHALUS.

In congenital cases, DR. F. DELAFIELD finds that very little can be done in the way of treatment, beyond supporting the nutrition of the child by suitable means, and correcting the disordered digestion.   By careful attention to these measures, in the milder forms of the disease, the child may improve.

In cases of acquired hydrocephalus, that is to say, where the lesion occurs secondarily, he finds that treatment is altogether useless.

Regarding *tapping*, DR. E. C. SEGUIN does not apprehend any special danger from the operation; this, he believes, will lie more in the hemorrhage than in the reaction, in these cases.   He advises, however, that if the hydrocephalus is due, as is sometimes the case, to a plugging of the aqueduct of Sylvius, tapping can only act as a mechanical relief. Also, if the disease affects the ventricular lining, he would hardly expect more than momentary relief from tapping.

CEREBRO–SPINAL MENINGITIS.

The treatment adopted by DR. F. DELAFIELD is very largely symptomatic. For the local lesion, he advises that *bleeding* should rarely be practiced in children, although in robust patients he sometimes abstracts blood from the temples, the neck, or along the upper part of the spine. This measure, he further cautions, is of service only during the first three or four days, and if the case is not seen until then, bleeding is contraindicated. He also applies cold to the head and back of the neck, by means of *ice-bags*, and in severe cases this is kept up continually for the first week. These measures are, he finds, the most efficient means of treatment, but insists that they are available only during the first stage of the disease.

For the restlessness and delirium, when excessive, he administers the bromides, or chloral, or hyoscy. amus. He frequently gives *potassium bromide gr. iii–x. every three hours*, but considers it more efficient when combined with one of the other drugs. Also, where the child cannot take the bromide alone, he usually prescribes the following :

℞. Potassii bromidi   .   .   . gr. iii–x.

Tinct. hyoscyami . . . m.xv–xl.

M. Sig. Dose, every three hours.

At times, he finds it necessary to resort to *opium*. Regarding *quinine*, he advises that it should be employed only as a matter of routine, as he believes that it has very little effect on the temperature. Where the fever runs high he prefers *cold* as an antipyretic; either in the form of affusions, or the cold or tepid bath, or occasional sponging which serves to render the patient more comfortable. *Stimulants* are also used by him in the later stages of bad cases, but, as a rule, he finds no indication for them during the course of the disease. For the vomiting, when it occurs in the early stage he disregards it altogether, unless very excessive, or continued beyond twenty-four or forty-eight hours; in such cases, he directs that small and frequent allowances of *iced milk and lime water* be administered, and the patient given small pieces of cracked ice to swallow. Should the vomiting still persist, he finds the following used with caution, very successful in checking it:

℞. Acid. hydrocyan.dil.. . . m. i–ii.

Sodii bicarbon. . . . gr. iii–v.

M. Sig. Dose, every three or four hours, to a child of eight to twelve years.

The constipation he relieves by an occasional mild laxative.

For the conjunctivitis, DR. DELAFIELD resorts to the continued application of *hot or cold cloths*, according as one or the other affords most relief to the patient, but, in either case, he insists that they be made hot or cold, not warm. When keratitis is present he uses hot cloths, together with the following:

℞. Atropiæ sulphat.   .   .   . gr. ii–iii.

Aquæ destil.   .   .   .   .   ℥ i.

M. Sig.   To be dropped into the eye, three times daily.

The inflammation of the choroid, he advises, cannot be effectively treated; all that can be done is to relieve the pain by a sufficient amount of opium.

For the general nutrition of the patient, this he maintains by a diet of milk, or milk and beef-tea, frequently administered.

DR. J. LEWIS SMITH places the child freely under the influence of *potassium bromide*, with ergot in the beginning, and the local use of cold in the shape of ice-bags, applied to the head and neck. He also resorts to the moderate and cautious employment of opiates, for the relief of pain and to procure rest. During the first two or three weeks he considers medicinal treatment of most importance. After this period has passed, however, his chief reliance is upon good nursing, and the use of a nutritious but non-stimulating diet.

In sudden attacks of severe cases, where the general symptoms are marked, opisthotonus prominent, and with jactitation, convulsions, etc., DR. F. A. BURRALL directs that a mustard bath be given and sinapisms applied to the neck, together with *sol. morph. sulph.* (*Magendie*) *m. iii-iv.* hypodermically, to a child of eight or ten years. This usually shortens the duration of the paroxysms and diminishes their intensity. Should, however, this condition persist or exaggerate in degree, with increasing stupor, he administers the *hot mustard bath three times daily*, and applies continuous strong *sinapisms* to the nape of the neck, with the effect usually of rendering the patient more quiet. These, together with the bath, are also continued as required. For the constipation, *hydrarg. chlor. mit. gr.* ⅛ is given every hour, until free fecal movement is established.

He also gives the following internally, and generally with beneficial effect:

℞. Potassii iodidi . . . gr. v.
Potassii bromidi . . . gr. x.
Aquæ . . . . . q. s.
M. Sig. Dose, every two hours.

In some instances, however, he has used up to ℨ i. of the bromide, with little or no relief whatever.

To control the restlessness at night and secure sleep, he gives *tinct. opii deodor.* ℥ ss. in divided doses. This, together with the bromide, is also freely administered during the day, when demanded.

For nourishment, a diet of milk is ordered, combined with stimulants, usually brandy, in the proportion of ℥ ss. to ℥ iv. of milk; this is given ad lib. and the brandy increased as required.

DR. W. H. DRAPER advises that the child be confined to bed, and kept perfectly quiet and at rest. As *rest* is considered to be thoroughly essential to the treatment, where there is much agitation exhibited he keeps the patient under the influence of opium, in sufficient quantity for this purpose. The rational measures employed by him consist in the internal administration of ergot and iodide of potassium. The *ergot* is given in large doses, in order to control the amount of blood carried to the spine; and, as a rule, he finds that the drug acts favorably in this disease, although in some instances ergot affords absolutely no benefit at all. Where the patient is unable to retain medicine by the mouth, he uses the hypodermic method. In conjunction with this *potassium iodide* is also administered, either by the stomach or rectum, with the object not only of controlling the inflammation, but also of removing the inflammatory products. In addition, the patient is usually allowed to take all the food wanted, causing no serious consequence.

These means of treatment are, in his experience, not always reliable, and in some instances are entirely negative. *e. g.* In certain cases he finds *salicylic acid in doses of gr. viii–x. every two hours*, very serviceable, on the plan of its utility in infectious diseases.

SPINA BIFIDA.

In young children where the general health is apparently good, while the tumor is present and small in circumference, Dr. L. A. Sayre believes that, as a rule, it is best to wait for nature to form a bony covering over the exposed parts, just as the fontanelles are closed in. In the meantime, should the tumor have reached any considerable proportions, he protects it artificially, and endeavors to assist ossification by the administration of internal remedies. The former he accomplishes by means of a *plaster jacket* carefully applied, and perfectly encasing the tumor; thus, also, guarding against any injury to the part. Or, a piece of steel or copper plate of appropriate shape, he finds, sometimes answers equally as well, provided it can be kept in exact position ; on account of this usual difficulty, however, he greatly prefers the plaster dressing. For internal treatment, he places the child on a course of *lime*, using the phosphate with a view of increasing the earthy phosphates in the system, and thus facilitating the further ossification of the spinal column.

After repeating the plaster application as occasion requires, DR. SAYRE sometimes resorts to aspiration with much satisfaction, and in many instances no inconvenience whatever is experienced during the operation, or subsequently, beyond some nausea and vertigo on rising in the morning. Where the tumor refills rapidly, he often finds it necessary to put in a permanent drain.

In infantile cases, DR. A. JACOBI sometimes places the child on the systematic administration of *ergot* and *iodide of potassium*, given in small doses, and continued for a considerable period of time.

DR. H. B. SANDS advises that although such measures are not entirely hopeless, yet operations for removal of the tumor are, as a rule, attended by disastrous results. When he resorts to any form of treatment, it is more frequently by the *injection of iodine*, or some such fluid.

[LISTER considers the following plan of treatment

the best at present known, and believes that incision
or antiseptic drainage, even with only one or two
horsehairs, do not offer a prospect of success. Where
the tumor is steadily increasing in size, he draws off
about an ounce of the fluid, and injects ℥ *i. of Morton's
iodo-glycerine solution.* If no effect is produced, in
about a week's time he repeats the operation, injecting ℥ iss. of the solution. As a rule, no bad symptoms follow, and, in favorable cases, under this
treatment he finds that the tumor gradually becomes
solid, and shrinks until it remains as a puckered fold
of skin only. Of late, however, in many instances
he recommends that no fluid whatever be drawn,
from danger of fatal consequences from loss of the
cerebro-spinal fluid.]

TYPHOID FEVER.

Dr. W. H. Draper directs his treatment, (a) to
maintaining the strength of the patient, and (b) to
guarding against the accidents which are liable to
occur.

Rest is considered absolutely essential. Any source
of irritation is avoided, and all disturbing thoughts

and muscular movements carefully prevented. This, he advises, should always be insisted upon, as patients often exhaust themselves greatly during the early stage of the disease, by being up and about as usual, and thereby lessen their chances for a good recovery.

For the *diet*, these patients, he directs, should not be over-fed. It should also be remembered, in feeding, for what object the food is given. When confined to bed, very little mental or mechanical work is done ; hence, the most important use for food is to repair the loss of heat resulting from fever.

If there is high fever with active delirium, more food is required than in low fever. Milk is believed by him to be the best kind of food for the great majority of cases. Sometimes, however, the patient cannot take a purely milk diet for any length of time ; in such instances, concentrated broths are given. These measures, judiciously employed, are, in many cases, all that he finds necessary.

In the severer forms of the disease, to control the high temperature is, he advises, the most important feature in the treatment, and must be accomplished by every possible means. If this is not done, the patient dies from simple exhaustion. For an anti-pyretic, *cold* is, in his experience, perhaps the most reliable , used in the form of the bath, pack, etc.

These, he finds, have the greatest power, and are therefore useful for the greatest emergencies. While, however, this is employed with benefit in the early part of the disease, yet in the later stages he considers it dangerous to the visceral lesions. The bath, he directs, should be kept by the side of the bed, to avoid any unnecessary moving of the patient. Its use is resorted to, and the process repeated as often as necessary, indicated by the fever; and in this way the temperature is continually kept near the normal standard. Where the bath is not practicable, he employs frequent sponging very effectually; either with warm or cool water, or with water and alcohol. This method is also less irritating to children, and is often quite successful in keeping the fever under control. This, however, he further assists by the administration of diuretics and diaphoretics, allowing the patient to drink freely of cold water, thereby cooling the blood, and favoring the excretion of matters which result from high temperature. *Cold water enemata* are also administered, too much covering avoided, and an abundance of cool and *fresh air* secured. This latter he finds exceedingly beneficial; indeed, in his experience, cases treated in the open air show a much larger proportion of recoveries.

Regarding the action of *quinine*, DR. DRAPER sometimes uses it in large doses, as an anti-pyretic,

and with good effect ; but its results have not proven
uniformly successful. When of service, he frequently
finds that it owes its power to an effect on the essential cause of the fever. Where, however, in this disease, the temperature makes rapid and marked ex-
cursions, he often gives quinine with great benefit.
But to attain this, he insists that the drug must be
administered in large doses ; small dosing does not
appear to have the slightest control in typhoid fever.
For hypodermic use, the following is employed:

R. Quiniæ sulph. . . . gr.clx.

Ac. hydrobrom. (Squibb) . ℨ i.

Spts. frumenti ad . . ℥ i.

M. Ft. sol.

Alcohol, he advises, acts as an antipyretic by fur-
nishing a food which is easily appropriated, and thus
supplies the force which the patient loses in heat,
and saves loss and exhaustion from combustion of
the tissues. The amount required varies in different
cases ; it is given until the desired effect is obtained.
In some children,

Vini xerici . . . . ℥ ii-iii.

daily, is found to be sufficient ; in others, brandy in
larger amount is necessary. By its stimulating effect
it also quells the delirium and controls the locomotor
disturbances. The heart's action becomes more
steady and firm, and the first sound, which may have

been feeble or absent, grows distinct. In all cases, however, he directs that its administration should be cautiously proceeded with, and considers it of the greatest importance to bear in mind the indications for its use, viz: Alcohol is demanded when the patient exhibits symptoms of marked nervous prostration, with subsultus, delirium at night, jactitation, with a position in bed indicating great physical debility, and with a dry tongue; all coming on with the increase of temperature. Under such circumstances, as a rule, it controls the delirium and the locomotor derangements, the patient sleeps quietly, the tongue becomes moist, and the whole aspect of the case is changed for the better. If, however, it increases the frequency of the pulse, if the delirium becomes more active, the tongue more dry, the skin hot and parched; if the motor spasms, tremulousness, subsultus, etc., are aggravated by alcohol, it is doing harm.

The following case, occurring in Dr. Draper's service at the New York Hospital, will serve very effectually to illustrate the power of alcohol as an antipyretic:

Boy, 13 years. Admitted Nov. 30. Four nights previous the patient was taken with severe pain in the abdomen, nausea, and vomiting, followed by high fever, with thirst, anorexia, prostration. etc. Pulse,

88; respiration, 24; temperature, 103°. Patient delirious.

Dec 1.—P. 118; R. 28; T. 105°. *Spts. vini gall.*
℥ viii. were administered during the day.

2nd.—T. 103°. Patient delirious. *Spts. vini gall.*
℥ xii.

3d.—T. 102°. Brandy diminished. In the evening: T. 106.° Brandy readministered as before.

4th.—T. 100.4°. 11 A.M., temperature nearly normal. Fever aborted; no diagnosis made.

For sustaining the heart power, besides the use of stimulants, digitalis is administered as indicated.

In the management of the accidents of typhoid fever, the most careful judgment is insisted upon.

Regarding leeches, blisters, etc., in his experience no benefit is derived from their application. Where perforation of a blood-vessel takes place, he enjoins *absolute repose,* and aims to control the diarrhœa as well as the patient's desire to defecate. To accomplish this, opium is given internally, and styptics administered per anum.

As to the value of *phenic acid* in typhoid fever, so strongly advocated by Dr. Déclat, this plan of treatment has been employed by him at the above hospital, but, from the results obtained, he does not consider it deserving of much credit.

During convalescence, DR. DRAPER advises that

the greatest care be exercised, from danger of relapse, perforation, and its serious consequence. Therefore rest and the most careful attention to the diet,
are always insisted upon.

DR. BEVERLY ROBINSON calls attention to the
great importance of the occurrence of sudden cardiac failure, in this disease.   He believes, from very
recent practical investigation, that at times, indeed
quite frequently, there is a sudden and considerable
dilatation of the cavities of the heart, more particularly of the right heart.   In such a condition coagula
may be formed, or, if *free and strong stimulation* be
at once resorted to, the period of danger may be
tided over, and only partial failure, with dangerous
but not fatal symptoms, occur.   To guard against
this peril, he advises that all unnecessary fatigue
should be absolutely avoided, during the course of
even mild cases of typhoid fever.   The patient should
not be allowed to rise unassisted in bed, or to sit up
but a few minutes on any occasion, and under no circumstances until convalescence is fully established.
Self-feeding, or any act which calls for the expendi-
ture of physical energy, and which can be prevented
by careful nursing, should never be permitted.   He
also resorts to the administration of heart tonics in
exceedingly moderate doses, from a comparatively
early stage of the disease, more especially if the

slightest evidence of failure of the heart power is exhibited. *Black coffee* is considered by him to be highly valuable as a stimulant, and he insists upon its use early. In addition he also urges the employment of liquid nourishment in the most concentrated form.

DR. F. DELAFIELD places the patient upon an exclusively milk diet, with confinement to bed. For an antipyretic he usually prefers cold, either in the form of the bath or the pack. Where the bowels are inclined to be loose, he administers :

℞. Bismuthi subnit. . . . . .

   Pepsini . . . . āā gr. iii–vi.

M. Sig. Dose, three times daily, or oftener if required.

If constipation prevails, *warm enemata* containing ox gall or castor oil, are used with good effect. In the later stages where this condition exists, he prescribes the *pil. aloes et myrrh. No. i–ii.* with much benefit. Stimulants are also given, usually whiskey or brandy according to indications.

In cases where dysentery supervenes, he directs that the bowels be freely opened with :

℞. Ol. ricini . . . . ʒ ii–ℨ ss.

   Tinct. opii . . . . gtt. iii–viii.

M. Sig. Dose, according to age.

Following this, he administers the *ipecac treatment.*

13

If much pain is present opium is given to control it. After the more active symptoms have subsided, he generally finds it of advantage to put the patient on the use of the following:

R. Tinct. opii deod. . . ℥ i. m.xx.
    Ol. ricini . . . . . . ℥ i.
    Mucilag. . . . . . . q.s.
    Aq. cinnamomi ad . . . ℥ iv.

M. Sig. ℥ ii–iv. three times daily, according to age.

On this plan the results are usually very satisfac tory.

In resorting to *cold as an antipyretic*, DR. W. H. THOMSON directs that the patient be placed in a bath of 75°–80°, and the water gradually cooled down to 65° or 60°; never lower, however. At the same time, cold affusions should be applied continuously to the head. Provided the fall is gradual, that is, one degree in six, four, or three minutes, the bath is continued until the temperature is reduced to 100°. If, however, the temperature falls one degree in two and one-half minutes, he stops the bath when the thermometer reaches 101°; for, as a rule, a still further reduction of 1° will occur after the bath is discontinued. Again, if the fall in temperature be one degree in two minutes, he removes the patient immediately, whatever the actual temperature may be at the time; for, in such cases, he advises, there is dan-

ger of the subsequent fall becoming uncontrollable,
reaching perhaps 97°, and the patient pass into collapse. Should such a condition occur at any time,
he at once orders the patient wrapped in hot blankets,
hot saucers applied to the epigastrium, and *stimulants*
freely administered. Where, from any cause, the bath
is impracticable, the cold pack is used ; always, however, with the same precautions.

In resorting to either of these measures, DR.
THOMSON directs that it be repeated often enough
to keep the temperature below the point of danger.
If necessary, he administers one every hour; usually,
however, two or three daily are sufficient, in which
case he advises that one should be so timed as to be
given just before the highest rise of the fever heat,
generally between two and three o'clock in the afternoon.

DR. A. H. SMITH uses the following method of
administering quinine by inunction, and finds it especially serviceable in children :

R. Quiniæ sulphat. . . . . ℈ i.
    Acid. oleic. pur. . . . . ℥ i.
    Ol. olivæ . . . . . ℥ ii.

Dissolve the quinia in the acid, with the aid of a
gentle heat, and add the oil. If properly prepared,
the solution will remain clear.

Sig. For inunction, to be well rubbed in.

This, he advises, may be applied once or twice daily to the entire surface, or oftener if required.

DR. AUSTIN FLINT directs that the child be placed in bed and on a milk diet, which is much preferred by him. As a rule, *alcohol* is also given in moderate amount, to sustain the strength of the patient, together with, in many instances, *spts. ammon. aromat.* For an antipyretic, he sometimes employs *quinine in large doses*, but usually relies more upon cold, in form of the bath, pack, or sponging. Of these, however, he places greatest reliance upon the *wet pack*, and finds this measure very generally successful, while a reduction of temperature does not always follow the use of the sponge bath. The cold bath, he advises, is not only a troublesome method, but it also frequently produces a nervous shock to the patient. The results, moreover, from the cold sheet, that is to say, by wrapping the body in a wet cloth and sprinkling with cold water, he finds equally as satisfactory in its effect.

In many cases occurring at BELLEVUE HOSPITAL, where the diarrhœa is excessive, the administration of *olei terebinthinæ, gtt. v–x. every three or four hours,* is found to be very effectual in controlling it. If hemorrhages occur and the patient passes into a condition of collapse, turpentine is given three or four times daily, in large doses, ice applied to the

right iliac fossa, and *sol. morph. sulph.* (*U. S.*) administered in sufficient amount to afford quiet, together
with *whiskey* ℥ *ii–iii. every half-hour*. Sometimes these
measures are successful in controlling the bleeding,
and the patient rallies. Or, if the hemorrhage recurs
and is more profuse, the administration of turpentine
is increased to every hour or every two hours, and
the following given :

℞. Ergotini   .   .   .   .   . gr. iii.

    Aquæ destil.   .   .   .   . ℥ i.

M. Sig. m. v–viii. hypodermically.

Morphia is also used to keep the patient quiet,
and hypodermics of *ether in doses of m. v–xv. once
or twice daily*, ordered, with whiskey as before. By
these means of treatment, repeated and persisted in
as occasion requires, the crisis is sometimes tided
over.

Where there is much insomnia, with restlessness,
the following will often cause the child to fall into a
quiet and refreshing sleep:

℞. Chloral. hydrat.   .   .   . gr. xl.

    Aquæ   .   .   .   .   .   . ℥ i.

M. Sig. A teaspoonful, repeated in one or two
hours, to a child of six to eight years.

Or, in many cases, the combination of potassium
bromide acts very favorably in controlling the cerebral excitement ; thus :

℞. Potass. bromidi   .   .   .   gr. x.
Chloral. hydrat.   .   .   .   gr. v.
Aq. anethi   .   .   .   .   . ℥ii.

M. Sig. Dose, every four to six hours, to a child of six years.

DR. A. L. LOOMIS directs that particular attention be paid to thorough disinfection, both of the stools and of the clothes and bedding. The room should also be well ventilated, and the temperature kept at about 60°. The *diet*, he advises, requires especial care. This should consist exclusively of fluids, and, in his experience, milk is the best article of food, to which lime water may be added, if necessary, to allay any irritability of the stomach. Regarding broths, gruels, etc., he not only objects to their use, but considers them often positively harmful. Fruits are also especially interdicted by him. In mild cases, where the fever does not range very high, he often finds that this is all that is needed, beyond an occasional or frequent sponging of the body with tepid or cold water.

In the more severe forms, his chief indications for treatment are to lower the temperature, maintain the heart's action, and support the nutrition of the patient. As an antipyretic in typhoid fever, he considers the use of *cold* in form of the bath, pack, etc., one of the most efficient. Ice-bags to the abdomen

and to the head, and *ice-water enemata* are also, in
certain cases, often found of marked value. Care is
advised, however, in the use of cold, and in all cases
he believes it unwise to resort to this measure after
the second week.

Quinine in large doses, is found by DR. LOOMIS to
render valuable service in the reduction of temperature, and in many instances this alone is employed.
In very severe cases, however, his custom is to use
both cold and quinine; that is to say, after the body
temperature has been lowered to 101°–102° by the
cold bath, he gives an antipyretic dose of quinine ;
thus maintaining the benefit derived from the bath,
and postponing the subsequent exacerbation of the
fever. The following formulæ are very highly
esteemed by him for hypodermic administration :

℞. Quiniæ disulph.   .   .   .   gr. l.
    Acid. sulphuric.   .   .   m. v.
    Acid. carbol. .   .   .   .   m. ii.
    Aquæ destil. .   .   .   .   ℥i.
M.
Or:
℞. Quiniæ sulphat. .   .   ʒi.
    Acid. hydrobrom.   .   .   .   ʒii.
    Aquæ destil. .   .   .   .   ʒvi.
M.
Of these, the latter has been more recently used

by him, and is much preferred. After the second week, however, should quinine alone, or, in certain cases, the use of the cold pack in conjunction, fail to control the fever, unless contraindicated by the pulse he often finds *powdered digitalis, in doses amounting to gr. v–x.* during the twenty-four hours, in combination with quinine, of excellent service.

For maintaining the heart's action, stimulants are given as indicated. But he advises that *alcohol* should be used cautiously, and always with discrimination. If its administration is of doubtful advisability, he directs that it be withheld until the indications for its use become stronger; and in all instances, he insists that the effects be carefully watched, and its use suspended if no benefit follows. At a period of crisis, however, free stimulation is often followed by the most gratifying results.

To meet the special symptoms and accidents occurring in the course of the disease, various means are employed according to circumstances. If the headache is very severe, *warm water fomentations* are usually found very beneficial. Should this fail, or the headache be followed by severe nervous phenomena, opium is given cautiously in small doses unless contraindicated by the pupil. Or small doses of *chloral*, repeated in two hours if necessary, often afford relief.

A moderate diarrhœa in the early stage, DR. LOOMIS believes may be let alone. Coming on during the third or fourth week, he endeavors to check it, placing the greatest reliance upon *opium* for this purpose. In his experience astringents, either alone or combined with opium, are of little or no service. Regarding the use of *cathartics* in the early part of the disease when constipation is present, he advises that they are very likely to do harm ; in his opinion they weaken the patient, increase the local intestinal lesions, and may cause perforation and peritonitis. For the tympanitis, he derives most benefit from the use of *turpentine stupes.* Intestinal hemorrhages in the early stage, when mild in character, require no treatment, as a rule ; moreover, he advises that profuse hemorrhages may.often be prevented by keeping the patient in bed. Coming on in the third or fourth week, however, or later, prompt measures are immediately resorted to ; *semi-narcotism* is induced by the use of opium, and perfect rest and quiet enjoined. Astringents he considers of questionable service. The occurrence of peritonitis is met by opium, as in cases of local peritonitis. Pulmonary complications are treated as usual.

During convalescence, DR. LOOMIS directs the strictest attention to the diet. The food is given frequently in small amounts, consisting of milk,

cream, broths, etc., avoiding solids, particularly fruits
and vegetables, and an advance to the ordinary articles of diet gradually and cautiously made.   If convalescence is retarded, he finds the use of tonics, cod
liver oil, and iron, very beneficial.   The following is
a favorite tonic mixture with him :

    ℞.   Quiniæ sulph.   .   .   .   gr. xxx.
       Ac. sulphuric. dil. .   .   .   .   q. s.
       Aquæ   .   .   .   .   .   .   ℥ ii.
       Tinct. ferri chlor. .   .   .   .   ℥ ss.
       Spts. chloroformi .   .   .   .   ʒ vi.
       Glycerinæ q. s. ad   .   .   .   ℥ iv.
    M.

.

SCARLET FEVER.

In mild cases, DR. ALONZO CLARK often finds that
medicines are not required at all, beyond a moderate
amount of champagne daily.   In severer forms of
the disease, for the throat affection, he uses calomel,
silver nitrate, etc.; but in children who are old
enough, he believes that gargling the throat frequently with cold water, or *cold carbonic acid water*
is, in many instances, the best means of treatment.

Where false membrane is present he prefers the spray of lime water, thrown into the child's mouth during inspiration, and allowing it to run down the fauces. If the disease is complicated, by diphtheria, he places· greatest reliance on the use of the *lime water spray*, continued and repeated as often as indicated; this measure is considered exceedingly efficacious. In regard to cold as an antipyretic, he recommends the *wet sheet* as very serviceable, and frequently resorts to it, although quinine is employed by him with much favor.

In those cases characterized by hemorrhagic eruption, he places the patient under the influence of *quinine*, combined with the free administration of the vegetable acids, and an abundance of nutritious food. Regarding the kidneys, if œdema occurs, he directs that the child be given a *hot bath*, and immediately afterward removed to a warm room and kept in bed, with sufficient covering to induce a constant, gentle perspiration day after day. This plan has been found by him to be the most successful method of treatment. DR. CLARK also advises that the bowels be kept free, not vigorously purged, but that a laxative be given as often as necessary, and an abundance of drink supplied, together with unirritating food. When indicated, an opiate or Dover's powder is also administered.

In the severer type of this disease, for the relief of the high temperature, DR. J. LEWIS SMITH resorts to the use of water, cold or lukewarm, applied in the form of sponging, the pack, or sometimes the full bath, whenever the thermometer registers above 103°. This he regards as one of the most valuable agents that can be employed, for reducing temperature and affording general relief to the patient. To control the nervous complications, he administers large and frequently repeated doses of *potassium bromide*, and usually very effectually. Where, however, this fails, he finds that the use of chloral hydrate, by the rectum, will almost surely prove successful. For this purpose he employs a small glass syringe, giving the following :

℞. Chloral. hydrat. . . . gr. v.
　　Aquæ . . . . . ℥ ii.

M. Sig. To a child of one to three years.

Regarding *inunction*, he considers it very serviceable, especially during the desquamative stage, preferring vaseline or carbolized applications. As to prophylaxis, he believes there is no remedy which can prevent the spread of the disease.

In the declining stage of scarlatina, DR. SMITH often derives great benefit from placing the child upon the use of the following :

℞. Ammon. carb. . . . . ℥ ss.

Ferri et ammon cit. . . . ℥ ss.

Syrupi . . . . . ℥ iv.

M. Sig. ℥ i–ii. every second or third hour.

If the disease is complicated by diphtheria, he ad
vises that stimulation is demanded ; in these cases,
he usually gives *brandy in doses of ℥ i. every half-hour*,
to a child of three years.

For prophylaxis, DR. A. L. LOOMIS employs those
measures necessary in other contagious fevers. The
patient is isolated, the room well ventilated, and a
thorough disinfection of both clothes and excrements
carefully attended to. Fresh air is, in his opinion,
of the greatest value as a prophylactic agent.

His medicinal treatment is chiefly symptomatic.
As an antipyretic, he considers *quinine in large doses*
most efficient. The cold bath is employed, however, where the temperature runs high, and the nervous phenomena present demand the adoption of
urgent measures. But in all instances, he directs
that the patient be sponged at frequent intervals ;
this affords comfort and, at the same time, relieves
the intense burning which is usually present, especially if a tepid saline water be used. When the process of desquamation begins, he advises that a warm
bath be administered once or twice daily, and *carbolized soap* used over the entire body. As regards the
use of stimulants, these, he finds, are sometimes in·

dicated even from the early stage of the disease, particularly in cases where marked nervous symptoms
are present, together with failure of the vital powers ;
and indeed, in certain instances, free stimulation constitutes almost his sole reliance.

He also directs that especial guard be made against
complications.   These are carefully watched for, and
met by suitable means.   For the throat, he finds
cold carbonic acid water, or pieces of cracked ice,
most serviceable ; particularly in the early stage.
Later on, when there is considerable infiltration present, he prefers the use of hot applications externally, covered with oil-silk, and combined with *steam
inhalations* and *warm gargles*, as of greatest benefit.
If ulcers are present, these are treated very satisfactorily with carbolic acid or *potassium chlorate*, in form
of the spray.   Anodyne applications, by means of
the spray, are also found to give much relief.

When evidences of kidney lesions are manifest,
DR. LOOMIS resorts to the immediate application of
cups, three or four in number, followed by the use of
hot fomentations.   Combined with these measures,
he also directs that the body be covered with flannel
and hot baths employed, together with the internal
administration of diuretics.   Of the latter he prefers *digitalis*.   Frequent draughts of water are also
given.   If these fail in reducing the anasarca, he then

administers *calomel* in combination with the diuretic, continuing this for two or three days, and often with the most gratifying result. If convulsions are imminent, or present, *opium* is given, affording relief to these symptoms as well as producing diaphoresis, which is sometimes of vital importance in this condition.

For a gargle, DR. G. M. LEFFERTS uses the following:

R. Glycerit. ac. carbol.   . .   ℥ i–ii.
Aquæ . . . . .   ℥ x.

M.

Or:

R. Acid. acetic. .   - .   ℥ iiss.
Glycerinæ . . . .   ℥ iii.
Aquæ . . . . .   ℥ x.

M.

DR. F. A. BURRALL finds the following very serviceable:

R. Acid. carbol. . . .   gr.xx.
Glycerinæ . . . .   ℥ i.
Sodii chloridi . . .   ℥ i.
Aquæ ad . . . .   ℥ viii.

M. Sig. Gargle.

DR. D. LEWIS has used the *infusion of digitalis* in the treatment of scarlet fever, in a large number of cases, and with markedly beneficial and uniformly

successful results. By this means, not only is the
frequency of the pulse lessened, the temperature
lowered, and its effects on the circulation of the kidneys exhibited, but he also finds that the tendency
to exudation in all glandular tissue is reduced to the
minimum, by its use. For these cases, he considers
the infusion of the best English leaves, the most reliable preparation. The tincture is, in his experience
unsatisfactory. In regard to the administration, he
directs that this be commenced at the earliest
possible stage of the disease, before those tissue
changes have occurred which it is intended to
prevent as well as cure. He usually gives the
following :

℞. Infus. digitalis . . . . ℥ i.

Sig. Dose, every four hours, to a child of five
years.

This, he finds, is a safe and generally an efficient
dose. Should, however, the pulse and temperature
be unaffected by it, he increases the amount to even
℥ ii. if necessary, but advises that its effects be carefully watched, and the dose reduced as soon as the
desired result is produced. Unless the pulse becomes less frequent than it should be, he continues
the administration two or three times daily, until the
end of the third week. This plan of treatment has,
in numerous instances, caused a decided fall in the

pulse and temperature within twenty-four hours, and
the most favorable results have followed.

For the throat affection, DR. LEWIS prescribes the
following, a favorite with him and usually very efficient:

℞. Potass. chlorat. .   .   .   .   ℨ i.
   Tinct. ferri chlor.   .   .   .   ℨ ii.
   Glycerinæ   .   .   .   .   ℥ i.
   Aquæ q.s. ad .   .   .   .   ℥ viii.
M. Sig. A teaspoonful every half-hour.

If exudation occurs in the fauces, he also uses this
mixture applied by the atomizer every twenty minutes. A *water sponge bath* is also given more or less
frequently, according to the degree of the fever, and
inunctions of olive oil applied over the entire body
at least twice daily, and continued until desquamation is complete. The diet he requires to be strictly
of milk, particularly on account of its diuretic effect.
In his experience, under this treatment, no case of
suppuration of cervical or other glands has occurred.

Regarding the nephritis, DR. A. JACOBI has succeeded in curing a number of cases by the following
method. He confines the child to bed and directs
that a *hot bath* be given twice daily, to produce free
diaphoresis; or sometimes the hot pack is used instead. In other instances, where the heart's action

14

is good, he employs the hypodermic injection of *pi-locarpine*, given two or three times in the twenty-four hours; thus inducing an abundant perspiration. In conjunction with this, he also administers *alcohol* in sufficient quantity to counteract the depression on the heart, caused by the pilocarpine. By this means, he occasionally gives as many as eight injections in twenty-four hours, always taking care to see that the child is well stimulated. This heroic plan is, however, not advised by him in all cases. In addition to the pilocarpine, he sometimes prescribes *gallic and tannic acid in doses of gr. v–xv., daily*, with very favorable effect. Toward the later stages, he finds the administration of *iron* often of great service, when there is much albumen in the urine. The cautious use of the salines is also resorted to, keeping the bowels gently open.

At the NEW YORK HOSPITAL, the following solution of pilocarpine is kept for hypodermic use:

℞. Pilocarpiæ mur.   .   .   .   gr. i.

Aq. destil. carbol.   .   .   .   m. l.

M.

DR. J. H. RIPLEY finds that in the majority of cases of scarlatinal nephritis the tendency is to recovery. His treatment varies somewhat according to the special symptoms, and the different complications arising in individual cases. In mild attacks, he directs that the child be put to bed at once, in a room the temperature of which is maintained at 70°–75°, and the atmosphere kept in a moistened condition. All exposure to cold is also to be carefully avoided. A mild laxative is given, usually a combination of *magnesia, senna and rhubarb*, keeping up a somewhat free action of the bowels; the use of hydragogues is, however, objected to by him. A diaphoretic is also administered, for which purpose he prefers the *liquor ammon. acetatis.* In addition to these measures, the diet is made simple and nutritious, alternating with milk, beef-tea, buttermilk, broths, etc.

In severe cases, where there is partial or complete suppression of the urine, he discountenances the use of *active diuretics* in the earlier stage of the disease, with the view of washing out the débris. In his experience the opposite result, with increased congestion of the already surcharged kidney, obtains, and he considers their employ worse than useless. When, however, the disease is subsiding, he then administers these remedies with marked benefit. His plan of treatment, and on which he places the greatest reliance, is to induce a free perspiration by means of the moist, *warm pack*, applied for at least two hours. This measure he finds highly efficacious, and believes it to be superior to any other means. In bad cases, he repeats the pack several times during the twenty-four hours, and, if necessary, every day for a number of days. In this respect, he cautions that great care must be exercised, to avoid any exposure to cold during the intervals. Should the child prove refractory, force is used; the hands and arms are tied and the coverings secured by safety pins, or other means; or, in many instances, he prefers to put the child under the influence of morphine, for this purpose. On removing the patient from the pack, the body is thoroughly rubbed with a coarse towel until the surface is dry. Regarding the *warm bath*, DR. RIPLEY objects to its use on account of the general tendency to cause exhaustion, if prolonged or frequently repeated. Moreover, he finds that many children are greatly afraid of it, and alarming syncope is often produced by this means. In older children, the *hot air bath* is occasionally substituted for the pack, although he considers it far less practicable.

Dry cupping is also resorted to, but he directs that the cups be removed before stagnation and rupture of the capillaries occur. As a rule, they are not allowed to remain on longer than five or ten minutes, and in delicate and nervous children he usually finds it better to administer *chloroform*, while applying them. In some cases, instead of cups, but more frequently following them, he applies *large hot poultices* over the parts, often mixing digitalis leaves with the meal, very efficaciously.

For internal medication, after perspiration is induced by these means, DR. RIPLEY keeps up a gentle moisture of the skin by administering the following:

R. Ext. jaborandi fl. . . gtt. v–xx.

Aquæ . . . . . . q.s.

M. Sig. Dose, every two hours, according to age.

This, he finds, is a much more serviceable way of giving the drug, than by larger doses and at longer intervals. If the temperature is high and the pulse full and strong, he gives *tinct. veratri viridis in small doses every two hours*, with marked benefit, and also with good effect on the kidneys. *Aconite* he finds less certain in its action, although not so likely to cause vomiting, and attended with less danger from prolonged use. If there is much restlessness present, morphia is employed.

Should excessive œdema occur, with general anasarca, and diaphoresis and catharsis prove unavailing, a few small *punctures* are made in the lower part of the legs, and often very successfully in causing removal of the fluid. As a rule, however, he rarely finds this procedure necessary in acute cases. Where pulmonary œdema occurs, he resorts to immediate dry cupping, together with the application of large hot poultices over the entire thorax.

In treating the uræmic symptoms most common in children, viz. : gastric irritation, painful and exhausting diarrhœa, extreme restlessness, etc., DR. RIPLEY cautions against the danger of inducing excessive purgation, especially by mercurials ; thus increasing the irritability of the stomach, impairing the nutrition, and causing loss of strength.   In his experience, producing more than two or three liquid passages daily is harmful.   For the gastric trouble, he advises the most careful attention to *diet*, giving the food in small quantities at a time; a teaspoonful every fifteen minutes, he finds, will be retained, when a tablespoonful every hour is vomited.   For this purpose, koumyss, milk and lime water, clam broth, or grated smoked-beef, are preferred by him.   If prostration is a prominent symptom, brandy and ice-water, or champagne, is given as indicated.   Or, in many cases, he considers the following almost a specific, prescribing it with greatest benefit :

℞.   Sol. morphiæ sulph. (U. S.)   .   gtt.v–xx.

Sig.   Dose, every three or four hours.

This is administered either hypodermically or by the stomach.   Or, sometimes a mixture of a few drops of chloroform with paregoric and syrup of acacia, does very well.   In others, he uses the following most satisfactorily :

℞.   Chloral. hydrat.   .   .   . gr.v–xv.

Aquæ . . . . . ℥ ss–i.

M. Sig. Dose, by the rectum.

To control the diarrhœa, he generally begins with a small purgative dose of calomel and compound jalap powder, and then employs a mixture of *opium, tannic acid and cinnamon water*, to hold it in check; avoiding, however, any tendency to produce constipation. If dysentery is present, he first administers a large enema of hot water, thoroughly washing out the gut and relieving the congestion, after which he gives the following:

℞. Tinct. opii . . . gtt.iii–x.

Aquæ ferv. . . . . ℥ ss.

M. Sig. Dose, as an enema, according to age.

This latter he repeats after each passage, the hot water injection being employed two or three times daily. Where the laudanum enema is not retained, opium in some form, either alone or with calomel, is employed.

For the restlessness, DR. RIPLEY usually finds this symptom yield promptly to *chloral hydrate* particularly when combined with small doses of morphine.

MEASLES.

For mild cases, DR. ALONZO CLARK finds that
hardly any treatment is necessary. The home
remedy, catnip tea, he considers about as good as
anything. In the severer forms of the disease, associated with complications, more urgent measures are
adopted. In the hemorrhagic variety, he administers *iron* and the *vegetable acids*, together with the
supporting effects of quinine and nutritious food.
Should gangrene occur, *quinine*, in his experience,
renders the greatest service, combined with the free
exhibition of stimulants. In the malignant form,
however, he finds medicinal remedies have little or
no influence over the disease. *Ergot*, he thinks, may
possibly afford some benefit, from its control over
the capillary circulation. *Cold*, applied both to the
body and to the head, he uses very effectually as an
antipyretic, when called for. Should anasarca occur,
he resorts to the same measures employed in his
treatment of scarlatina. For the ophthalmia which
sometimes follows an attack of measles, DR. CLARK
recommends the application around the eyes of an
ointment of veratria, thus:

℞. Veratriæ   .   .   .   .   gr. viii.

Adipis prep.   .   .   .   .   ℥ i.

Ol. olivæ   .   .   .   .   .   ℥ ss.

M.

This he uses with the utmost satisfaction.

As prophylactic measures, DR. A. L. LOOMIS directs that the child be placed in a separate room, and an abundance of fresh air obtained; together with disinfection of the clothing, attention to cleanliness, etc., and good hygiene. In mild cases, very little is done in the way of treatment. The room is darkened, good ventilation secured, avoiding draughts and all exposure, and the patient placed on a milk diet with occasional broths. The fever, where it is slight in degree, is disregarded, or controlled by frequent tepid sponging, which also allays the itching and burning of the skin. Where there is much thirst, this is relieved by permitting the child to drink freely of cold water, in small amounts. In these cases, as a rule, he objects to the use of *stimulants* in the early stage, and advises that very serious results may follow their administration. When, however, prostration is marked, and with the typhoid condition present, the strength of the patient must be supported, and stimulants are often required from the onset of the disease. If there is considerable restlessness, *opium* is given in small doses; usually Dover's powder.

Where the disease runs a severe course, and the fever calls for more active measures, DR. LOOMIS places most reliance upon *quinine*, as an antipyretic. He uses the *sponge bath*, however, very effectually, but objects to the cold bath from fear of pulmonary complications. These, he advises, should in all cases be especially guarded against ; and when bronchitis is present, or as soon as laryngeal symptoms are manifest, and respiration interfered with, he immediately resorts to *inhalations of vapor*, and continues them so long as evidence of any obstruction to the breathing exists. The importance of this measure, he believes, cannot be over-estimated.

The following is a most excellent cough mixture, and is used very extensively with children at BELLEVUE HOSPITAL (also at the INFANTS' HOSPITAL):

R̸. Tinct. opii camph. . . .
Spts. ammon. arom. . . . āā ℥ i.
Ext. ipecac. fl. . . . . ℨ ss.
Syr. pruni virg. . . . . ℥ i.
Aquæ q.s. ad . . . . ℥ viii.
M. Sig. A teaspoonful.

In cases of so-called German Measles, or Epidemic Roseola, his treatment consists mainly in careful attention to the diet, etc., and the avoidance of exposure. Any catarrhal affections which are apt

to arise in the course of the disease, are carefully
guarded against, and when occurring, receive prompt
attention. For the itching, which is sometimes intolerable, he finds sponging with tepid water very
efficient, as well as in relieving the febrile symptoms.
In certain cases, he also prescribes tonics, *iron*, and
cod liver oil when indicated, with good effect, but
usually he finds that the disease runs a mild course,
and recovers with no untoward symptoms.

VARIOLA.

(SMALL POX.)

DR. J. N. MCCHESNEY (late Attending Physician
to the Hospital for Contagious Diseases) finds that
mild, discrete cases require little or no treatment,
except the usual attention to hygiene and diet.
This latter, in the early period of the disease, is
made simple, consisting of milk, rice, cornstarch,
etc. In these patients, when contact with other
persons can be prevented, exercise in the open air is
also found beneficial.

During the first few days, in the early stage, he
administers cooling acidulated drinks for the fever,

together with sponging the body with cool or tepid water; this also serves to render the patient more comfortable. *Quinine* he considers of little or no use as an antipyretic, and finds, moreover, that it is rejected, as a rule, and only adds to the irritability of the stomach.   If constipation is present, an enema or mild laxative is given, avoiding active catharsis.   For this purpose, the *citrate of magnesia* is found very agreeable and efficient; or sometimes seidlitz powders, or one of the aperient waters, are used.   These also tend to relieve the nausea and vomiting.   Where diarrhœa is prominent, mistura cretæ is ordered, either alone or with one of the vegetable astringent tinctures.   Or, in many cases, the following is used by him with the greatest satisfaction :

  ℞. Ext. coto fl. . . . . m. x.

    Syrupi aromat. . . . q. s.

  M. Sig. Dose, every half hour.

This he gives until three doses have been taken, which is usually found sufficient.   The nausea and vomiting, however, is sometimes very persistent and distressing.   To relieve this symptom, in some instances small lumps of ice, swallowed whole, are of service.   In others, lime water, or *carbonic acid water* in small quantities is most efficacious.   Or, morphia hypodermically over the region of the stomach is often used with the greatest success. *Counter-irritation*, in the shape of mustard poultices, though usually effectual, is objected to on account of inducing vesication. For the headache, he applies cold cloths; or, occasionally hot cloths, as hot as can be borne, are more serviceable. If there is much restlessness and delirium, he gives the *bromides combined with chloral* in an aromatic vehicle, with good effect; or, if necessary, morphia is used. To relieve the intense pains in the back, various means are employed. Sometimes the sponging and bathing resorted to for the temperature, is found beneficial. Or the application of a sponge wrung out of hot water, repeated as soon as it cools, often affords relief. At times, however, *hypodermics of morphia* are needed. With the appearance of the eruption, he finds that for a few days these mild cases call for no treatment whatever; as the symptoms are greatly diminished in intensity, and in some instances are altogether absent.

In the more severe attacks, to relieve the œdema of the face and eyelids DR. McCHESNEY applies *hot water compresses* constantly, changing them at frequent intervals. The smarting pain, generally present over the entire surface of the body, is treated by the application of cold vaseline, which is much preferred by him to the numerous ointments, creams, etc. For the throat symptoms, various gargles are at times employed; *e. g.*, flax-seed tea, solutions of potassium chlorate, alum, and borax. In his experience, however, *bromo-chloralum* has given the greatest satisfaction, usually affording quicker and more lasting relief. Small pieces of cracked ice, held in contact with the mucous membrane, are also very grateful. Externally, he uses hot moist applications with benefit; together with *steam inhalations*, and spraying the throat with astringent solutions, which often give much comfort. For the conjunctivitis, in most cases a weak solution of alum answers very well. When, however, there is also pain and photophobia he employs the following very effectually:

R. Atropiæ sulph. . . . gr. ii–iii.

Aquæ . . . . . . ℥ i.

M. Sig. To be dropped into the eye.

Where the pain from distension of the vesicles on the hands and feet is intense, he finds that soaking the part in hot water for fifteen or twenty minutes, followed by *puncture* of the vesicles and escape of their contents, affords most relief. In confluent cases, when the bullæ break, marked benefit is derived from dusting the raw surface with a powder of *bismuth and zinc oxide;* or sometimes lycopodium is used for this purpose.

In cases marked by the early appearance of great prostration and restlessness, stimulants are freely administered, combined with nutritious food. When extensive suppuration is evident, the strength of the patient is supported by the early use of *stimulants* resorted to, according to indications, and continued until convalescence is fully established.

For an antipyretic in the stage of secondary fever, DR. MCCHESNEY has tried quinine in large and frequent doses, but with poor results; hence, he greatly prefers *cold*, in the form of sponging, or the wet pack, and places the utmost reliancé upon its use. This, he finds, not only controls the temperature, but also the delirium and restlessness. In administering it, he always begins with water that is slightly tepid, and continues using it until the temperature falls to about the normal. In weak patients, the addition of alcohol to the water is often made with advantage. If anodynes are required, he finds the *bromides* or *chloral* very serviceable. Caution is advised, however, in their use, particularly in cases where there is an abundant expectoration of viscid mucus; as, during sleep, the accumulated secretions may pass into the air-passages in sufficient amount to cause asphyxia. During this stage, the *diet* is made of the most nutritious character, consisting of milk, beef-tea, broths, eggs, etc. Stimulants are also given, alone, or in the form of milk punch and the like.

When the pustules rupture, to relieve the burning and itching of the skin, he finds carbolized baths most effectual, followed by the free use of vaseline. To correct the offensive odor, he much prefers a solution of *potassium permanganate,* or bromo-chloralum, to carbolic acid. Occurring in the hot weather, he advises that great care be taken in preventing the flies from depositing their eggs in the pustules, and which, when developed, may cause serious results. For this purpose he finds the free application of a strong solution of permanganate of potassium the best method, not only of destroying them, but of preventing their return.

During the period of desquamation, especially if suppuration has been abundant, he administers *tonics,* iron, quinine, and the bitters, combined with nutritious food and the use of stimulants. At this time, frequent *warm baths* are found very agreeable, as well as assisting in the removal of the crusts. In cases where the scabs about the face and nose are unusually adherent, and tend to prolong and increase the ulceration, he directs that they be softened with hot water, or vaseline, and removed. After which, he makes direct application to the bottom of the ulcer of the following:

℞. Iodoformi . . . . gr. xxx.

Bals. tolutan. . . . . ʒi.

M.

By this method, he has obtained the most uniformly gratifying results. In regard to the prevention of pitting, DR. MCCHESNEY has tried every means recommended by various authorities for this object ; but it is his experience that if the pustules are superficial, the pitting will be slight, while if the true skin is involved, this will occur in spite of every endeavor.

During convalescence the hygienic surroundings receive the most careful consideration ; plenty of *fresh air* is secured, at the same time avoiding any exposure, from the risk of pulmonary complications.

The following, from BELLEVUE HOSPITAL, will be found very useful if broncho-pneumonia sets in :

℞. Ammon. carb. . . . . ʒ ss.

Syr. senegæ . . . ʒ iv.

Syr. ipecac. . . . . ʒ ii.

Syr. tolutan. . . . . ʒ iv.

Ext. glycyrrh. . . . . ʒ ss.

Aq. cinnam. q. s. ad . . . ℥ iv.

M. Dose : A teaspoonful.

DR. ALONZO CLARK directs that the patient be placed in a large, well-ventilated room provided with an open fire place, and from which the carpets and all unnecessary furniture should be removed. The

15

strictest attention should also be given to disinfection. His medicinal treatment is in the main symptomatic. He advises that in very many cases of
small-pox, where the tendency is to recover, but
little treatment is necessary. If there is much exhaustion, he administers stimulants as in typhoid
fever. Where a wide erysipelatous inflammation is
present, he directs the room be kept cool, so that the
coldness of the air may modify the action of the
inflammatory disease. Should the patient exhibit
great restlessness, opium is given in some form, to
afford quiet. If œdema glottidis occurs, *scarification*
is resorted to.

Regarding the subsequent cicatrization, he finds
that if early in the disease the vessels be pricked
with a lancet, and kept empty, such a condition may
sometimes be avoided. Or, the application of the
old *mercurial plaster* to the face, he occasionally uses
with more or less success in preventing the pitting.
Another method recommended by DR. CLARK, is by
the use of a mixture of collodion and animal black,
applied with a brush over the face.

The following is the formula used at the Children's Hospital, Paris, applied in the form of a plaster,
and is considered to be quite successful to prevent
the pitting:

℞. Ung. hydrarg. . . . part. xxv.

Ceræ flavæ . . . . part. x.

M. Picis nig. . . . . part. vi.

In the early stage of the disease, DR. A. L. LOOMIS directs his treatment mainly to the attendant symptoms. If the thermometer shows any marked degree of fever, he controls it by the use of cold, in the form of the bath or pack; or by the administration of *quinine in large doses*. When there is much vomiting, iced carbonic acid water is given with good effect; or if with this symptom there is associated considerable restlessness, opium is resorted to. Where constipation is present, *cold water enemata* are found very serviceable, as well as having a cooling effect on the blood. For the headache, when severe, cold in the shape of *ice-bags*, or compresses, is applied with benefit.

With the appearance of the eruption, the means resorted to by him depend upon the character of this stage. Mild cases of discrete form, he finds, call for little or nothing in the way of treatment. Should, however, the eruption be slow in development, or late, with high temperature, he finds that a *warm bath* for a quarter of an hour or longer often hastens its development. When desiccation is reached he directs that warm baths, followed by oiling of the body, should be administered every one or two days. In the severer forms, where there is great depression.

delirium, the typhoid state, etc., and especially dur·
ing the suppurative stage, *stimulants* are freely administered and often with the most gratifying effect.
Where the delirium is very marked, opium is also
given in combination.

In regard to the pitting, DR. LOOMIS has tried
a number of methods to prevent its occurrence, but
finds that the best results obtain from opening the
vesicle before it becomes a pustule, and dressing the
part simply with cold water.

DR. W. H. THOMSON recommends the following
as a very excellent tonic formula, when indicated :

  ℞.   Ferri et ammon. cit.   .   .   .   ʒ i.

        Ammon. carb.   .   .   . gr. xxx.

        Tinct. gentian. co.   .   .

        Tinct. quassiæ   .   .   . āā ʒ ii.

        Syrupi   .   .   .   .   . ʒ iss.

        Aquæ q. s. ad   .   .   .   ʒ viii.

  M.   Sig.   One to two teaspoonfuls.

INTERMITTENT FEVER.

DR. AUSTIN FLINT gives quinine at any stage of
the disease, until slight signs of cinchonism are de-

tected. He advises beginning with small doses, to ascertain the tolerance for the drug ; usually thus :

℞. Quiniæ sulph. . . . gr. i–iii.

Sig. Dose, every four hours.

Having discovered this, he continues the administration in full doses until the paroxysm no longer occurs, and in smaller doses for a long time afterward. He generally gives it by the mouth, or, in other cases, by the rectum in double the quantity by means of enemata. *Iron* is also employed for the anæmia attending the disease. For the enlarged spleen, he finds that nothing so speedily diminishes the size as quinia.

DR. ALONZO CLARK considers the following combination a most excellent one, and of peculiar value in many cases :

℞. Quiniæ sulph. . . . gr. xx.

Pulv. capsici . . . . gr. vi.

Pulv. opii . . . . . gr. i.

M.

This he has given in very obstinate cases of long standing, where the spleen has been much enlarged, and with marked benefit. In all cases, however, he administers quinine daily for some time, whether signs of intermittent fever present themselves or not. Where cachexia is also more or less prominent, he administers iron, usually in form of the carbonate with much satisfaction. For this purpose the chocolate iron lozenges, which contain *ferri protocarbon. gr. iiss.*, are greatly preferred by him. The advantage of this form of iron preparation, he advises, is that the chocolate renders the iron quite tasteless, and very agreeable to take.

DR. F. A. BURRALL directs attention to a condition occurring in children of from one to two years, sometimes older, which he finds is not infrequently mistaken for malaria, but which is regarded by him as acute blood poisoning due to defective hepatic secretion.   [In the milder forms, he advises, the child suddenly or gradually, while in usual health, becomes indisposed, pale or perhaps faintly yellowish, has a dry and somewhat warmer skin than natural, and an irritable stomach with constipation, or scanty light-colored stools.   If this condition is not arrested it tends to further development, when the surface becomes decidedly yellow and dry, the pulse frequent and the temperature elevated, with headache more or less severe, grating of the teeth, and perhaps delirium.   The appetite is sometimes voracious, sometimes lost.   The irritability of the stomach varies from nausea to persistent, and occasionally apparently uncontrollable, vomiting.   Repeated chills usually occur, and in some cases sore throat is present.]   In these attacks, his indication for treatment is to promote the excretions. To accomplish this, he gives:

℞. Hydrarg. chlor. mit. . . gr. ¼–½.

. Sig. Dose, every hour, placed dry upon the tongue.

The administration is then repeated until four or five doses have been taken, or the bowels begin to act. This plan, he finds, is usually attended with marked success.

DR. J. C. PETERS has also met with a number of cases of supposed malaria, of the class described by DR. BURRALL and has likewise obtained good results from the use of calomel. In his experience, however, other remedies answer an equally good, and perhaps better, purpose. The following is his favorite formula for these children :

℞. Tinct. aloes (U. S.) . . part. iv.

Ext. glycyrrh. fl. . . pars i.

M. Sig. m. xv–xxx–ʒ i. according to age.

In many so called "mixed" cases of malaria, which do not always yield promptly to quinine, he finds that by a combination with one of these, the quinia acts more favorably than when given without an adjuvant.

Certain cases of malaria occurring in older children, have also been noticed at the NEW YORK HOSPITAL, in which vesical irritability with nocturnal inconti-

nence of urine, pubic pain, etc., have been present;
and in some instances even simulating vesical calcu-
lus, from which, moreover, careful judgment is often
necessary to diagnosticate.   In these cases, the
patient is placed on the following treatment with
very gratifying results.   Laxative pills are ordered,
one to be taken three or four nights in succession,
together with *quiniæ sulph. gr. iii–v. two or three
times daily.*   In conjunction with this, a pill containing *ext. belladonnæ gr.* ⅛ is also administered three
times daily.

In cases where anæmia is very marked, the following is often found of great benefit :

℞.   Ferri et potass. tart.   .   .   gr. v.

Liq. potass. arsenit.   .   .   m. ii.

Potass. bicarbon.   .   .   .   gr. x.

Tinct. nucis vomicæ   .   .   m. v.

Aquæ ad   .   .   .   .   . ℥ i.

M.   Sig. To be taken in a wineglass of water, before eating.

During the paroxysm, Dr. A. L. Loomis directs
that the patient be put to bed, and, in the cold stage,
well covered with blankets, bottles of hot water applied to the surface of the body, and hot drinks
freely supplied.   As the hot stage comes on, these
are gradually removed, and cold drinks administered.
The nausea and vomiting he usually relieves by

means of opium. In the sweating stage, no treatment is called for. During the interval, he gives a
moderately large dose of quinine at the close of the
third stage, and double the quantity two hours before
the occurrence of the next paroxysm. Where the
stomach is too irritable the hypodermic method is
used; for this purpose, the following combination is
greatly preferred by him:

℞. Quiniæ sulphat. . . . . ℥ i.
    Acid. hydrobrom.   .   .   . ℥ ii.
    Aquæ distill.   .   .   .   . ℥ vi.
M.

In very nervous cases he often gives opium in
combination with good effect. A mild state of
cinchonism is then kept up for several days; after
which he directs that the patient should be seen one
month after the first paroxysm, as the chills are very
liable to return at this period, and cinchonism should
be again produced. When quinia alone fails, because
of an hyperæmic condition of the liver and spleen,
he finds that *calomel in full doses*, combined with
the former remedy, will often prove of greatest efficacy. Arsenic he considers of little or no service.

In those cases where the infection has assumed a
chronic form, DR. LOOMIS advises that the patient
be removed to a warm, dry, non-malarial region, and
all exposure carefully avoided. *Tonics* he finds very

useful. Iron is given with the quinine, and often proves exceedingly serviceable when anæmia is prominent. Where the liver and spleen show much enlargement, he considers the iodide of iron in combination with cod liver oil particularly valuable. He also advises that the bowels receive careful attention, and constipation, when present, relieved by rhubarb or aloes. *Arsenic* is often found of benefit, especially in cases where the above means have failed. The administration of the drug, he advises, must be watched with great care, and immediately stopped on the appearance of any constitutional effects. Hygienic and dietetic regulations also receive proper care. A nutritious diet is considered by him to be of the highest importance, in these cases.

At BELLEVUE HOSPITAL the following means are often employed, in cases of older children, to prevent the chill of malaria. In a large number of instances, chloroform and whiskey are often found very efficacious; given thus:

℞. Spts. chloroformi . .

　　Spts. frumenti . . . āā ℥ i–ii.

M. Sig. Dose.

Or, the following is administered, and not infrequently entirely prevents the occurrence of the chill :

℞. Amyli nitriti . . . gtt. iii–iv.

Sig. To be inhaled every half-hour, from a cloth or sponge.

In other cases, *pilocarpine in doses of gr.* $\frac{1}{12}$-$\frac{1}{8}$ *hypodermically* is often of excellent service. For the administration of quinine by hypodermic injection, the following formula is very generally employed at this hospital:

℞. Quiniæ sulphat. . . . gr. lxxx.
Aquæ . . . . . ℥ i.
Acid. sulphuric. dil. . . . q. s.
Heat to boiling et add.
Acid. carbol. . . . . gr. v.

M.

TYPHUS FEVER.

Dr. Austin Flint places the patient on a diet of milk, eggs, broths, etc., advising, moreover, that these should not be given at too short intervals. The system is also supported by stimulation, particularly where any undue depression is present. Or, if there is any doubt as to the advisability of its administration, he directs that the alcohol be given and its effects closely watched. *Brandy* is usually preferred by him, in small doses at intervals of one or two hours, thus determining the amount required daily; in all cases, however, its toxical effect should be carefully avoided.

For the high temperature, he places greatest reliance upon *cold* as an antipyretic, although in some instances quinine in large doses is found very serviceable; or, at times, aconite or veratrum viride is used. In the application of cold he considers the wet sheet the best method for continued use. Where the bath is employed, if it gives rise to no unpleasant symptoms, he directs that it be continued until the temperature is reduced; always bearing in mind that a fall of one degree or more may be expected after removal from the bath. When the temperature again rises above 103°, the process is repeated. Regarding this plan of treatment, he finds that unless it be systematically carried out, its merits cannot be fairly judged. The sponge bath is also used very effectually, besides affording comfort to the patient.

For the headache, cold, by means of the douche or the wet napkin, applied to the head after having cut the hair close, is found of marked benefit. This is also of service in quieting the delirium, when active, combined with the internal use of *opium,* either alone or with small doses of tartar emetic. Where the delirium is mild in character, no especial treatment is required. If insomnia is a prominent symptom, this he controls by the use of the *bromides ;* or sometimes opium is necessary.

To relieve the nausea and vomiting, DR. FLINT advises that the greatest care be exercised in the administration of the food, as he frequently finds this symptom due entirely to improper management of the diet. In other instances bismuth and opium powders are given with good effect. This also serves to control the diarrhœa, if present. The following is used at BELLEVUE HOSPITAL :

R. Bismuthi subnit. . . . gr. iv.

Pulv. ipecac. co. . . . gr. i.

M. Sig. Dose.

Where, however, there is diarrhœa of a mild form, it may be let alone ; in his opinion, more than three or four passages daily would indicate the resort to measures for suppressing it, when astringents, turpentine, etc., may be needed. For the constipation which is usually present, enemata are employed.

In addition to these means of treatment, the observance of good hygiene, cleanliness, plenty of fresh air, etc., receive the most careful attention.

The following is often found to be an exceedingly serviceable prescription in the condition of insomnia and delirium, and is highly recommended by several authorities who claim for it magical effects, the patient waking refreshed and rational:

℞. Liq. opii sed. (Battley) . . ℨ i.
Antimon. et potass. tart. . . gr. i.
Aquæ camphoræ . . . ℥ vi.

M. Sig. A dessertspoonful every hour, until sleep is induced.

Regarding prophylaxis in the treatment of this disease, DR. A. L. LOOMIS advises the strictest regulations of quarantine, disinfection, ventilation, and careful attention to cleanliness, etc. He directs that the patient be placed, if possible, in a large, airy room, free of all materials likely to retain infection, and an abundance of fresh air secured at all times; this he considers a factor of the greatest importance, and believes that mild cases require very little else in the way of treatment. The *diet* also requires the most careful attention, as regards the selection of proper food, its administration, and the avoidance of all over-feeding. Force may even be necessary for the sustenance of the patient; or the stomach tube passed through the nares, is often of great service for this purpose. Milk he considers the best food for these patients, given frequently and at regular intervals; or milk with yolk of eggs may be used when desired.

For the reduction of the temperature, he places most reliance upon *quinine*, as a rule, although the cold bath is often used, particularly in the earlier stages of the disease where the exacerbations are more rapid. If necessary this is also assisted by a full dose of quinine, administered about twenty minutes after removal from the bath. *Ice-bags* to the head are found a serviceable adjunct to the bath, especially where there is much pain in the head, or if active delirium is present during its administration. After the first week, however, he usually finds that quinine alone is a sufficient antipyretic, particularly if the fever has previously been kept below 103°.

Regarding the use of *alcoholic stimulants*, DR. LOOMIS does not favor them in all cases, and especially in young subjects. To control the pulse, however, and support the heart's action, stimulants are given as indicated and with much benefit. Where prostration is extreme, the judicious use of alcohol is, in many instances, followed by very satisfactory results. In other cases, where frequency of the pulse, occurring in the early stage of the disease, is due to a failing heart, digitalis is given and often with excellent effect ; as follows:

℞. Infus. digitalis . . . ℥ ii–iv.

Sig. This amount in twenty-four hours.

For the intense pains in the head, cold in the

shape of ice-bags renders marked service; or when associated with photophobia, a small *blister* applied to the back of the neck, is found very efficacious. The insomnia which is frequently present is also usually relieved by the use of cold applied to the head ; if it persists, however, in spite of this measure, opium is given to control it.   When other marked nervous phenomena are also present, he finds the careful administration of *chloral hydrate* very beneficial.   If stupor is marked, this is met by the local resort to stimulating applications, combined with the internal use of the diffusible stimulants ; of which he prefers *musk* and *camphor*, or in some instances coffee.

During convalescence, DR. LOOMIS advises great care in avoiding all over exertion, exposure, etc. The appetite must also be held within bounds.   At this time, tonics, iron and quinine, and the mineral acids are often of service, especially if there is feeble heart action.   For this purpose, the following, one of Dr. Loomis' favorite tonic formulas, may be found of marked benefit:

℞. Sol. quiniæ sulph. (gr. xv–ℨ i.)   .   ℥ ii.
    Tinct. ferri chlor.   .   .   .   ℥ ss.
    Spts. chloroformi   .   .   .   ℨ vi.
    Glycerinæ q. s. ad   .   .   .   ℥ iv.
M.

OTITIS.

(EAR-ACHE.)

For the pain, which DR. O. D. POMEROY believes should receive primary consideration, he finds the application of one or two *leeches* to the part, the best plan of treatment. If, as occasionally happens, the leech aggravates the pain, morphia is cautiously used, which acts as a true anti-phlogistic as well as anodyne. He then repeats the bleeding, if necessary, so long as the pain or sensation of fullness continues; care being taken, however, not to produce exhaustion. In many cases, he advises, milder measures suffice; such as the application of *dry heat*, by means of a rubber bag, or a bottle, of hot water, the temperature being adjusted to the comfort of the patient. Water either too hot or too cold, he finds, will only serve to increase the pain. Hot salt bags are also used very effectually, or sometimes a roasted onion acts well. Regarding moist applications, in his experience although they afford relief, yet, if continued too long, otorrhœa and other serious results may follow. A little *paregoric* on cotton is also serviceable, or at times morphia is cautiously employed. In other cases, he finds that a bit of cot-

16

ton with black pepper wrapped in it, will warm the parts and allay the pain. (This should first be tried on an adult, so as to avoid excessive burning.) The following is also regarded by him as very efficacious:

℞. Atropiæ sulph. . . . gr. iv.

Aquæ . . . . ,.. ℥ i.

M. Sig. To be dropped into the eye.

Steam, or the vapor of chloroform, blown into the ear, are also employed with good result.

Regarding *puncturing* of the membrane, if there is a collection of the secretions in the tympanum, DR. POMEROY considers this operation more appropriate for the relief of pain, than leeching. Where no discharge follows the puncture, he then resorts to inflation; and by inclining the head to the affected side, all the discharge may in this way be blown out. This is repeated every day or two until the accumulation has finally ceased. After the pain is relieved, he directs that cotton be kept in the ear, removing it as often as it becomes moistened. The ear is also kept clean by *injections* and syringing, very gently (so as not to cause a return of the throbbing pain), with a warm saline solution (℥ i–Oi.), and carefully wiped out and dried once or twice daily. After two or three days he inflates the drum, and if the discharge does not diminish in from four to six days, he then resorts to astringents. For this purpose the following is often preferred by him :

℞. Plumbi acetat. . . . gr. ii–v.

Aquæ . . . . . ℥ i.

M. To be poured into the ear twice daily, after syringing.

Or, sometimes *argenti nitrat.* of the same strength is used. If a disinfectant is needed, the following may be employed instead of the above :

℞. Acidi carbolici . . . . ʒ i.

Aquæ . . . . . ℥ xvi.

M.

But, in any case, he advises that whatever astringent be used, none should cause excessive pain or make the ear throb afterward.

This plan of treatment DR. POMEROY usually finds quite successful. Where, however, the discharge does not disappear, he then fills the canal with *boracic acid*, finely powdered, well packed in by means of cotton (using a holder), and permits it to remain until it becomes moistened by the discharge, when it is removed by syringing and again renewed.

DR. A. JACOBI finds that in many severe forms of ear-ache, where the trouble is probably caused by a catarrhal affection of the Eustachian tube, by closing the mouths of infants and children and simply blow·

OTITIS.

ing into the nose, is often a very valuable method of affording relief to the pain.

DR. R. F. WEIR uses *morphine and atropine* very effectually to control the pain, together with injections of hot water, that is to say, by pouring the water into the ear, not syringing it; and in many instances this latter measure alone suffices, the child falling asleep during the procedure. Leeches he objects to. Regarding *paracentesis*, he finds that especially in children the external canal is often swollen and painful, thus rendering the operation difficult and impracticable; hence, he advises that, as a rule, it is best to let the drum alone.

DR. A. A. SMITH is in the habit of using *warm douches*, together with the application of heat in the form of hot salt bags. In some cases, however, he finds it necessary to give a full dose of an anodyne. He has also obtained good results from blowing *chloroform vapor* into the ear.

DR. BEVERLY ROBINSON has derived benefit from the application of leeches, when all other means had failed. He also finds much relief afforded by the use of the continuous douche. In other instances, he has used *salicylate of soda in large doses* very successfully in allaying the pain.

DR. C. R. AGNEW considers the warm douche and opium of the greatest value.

In all cases, DR. S. SEXTON regards *rest and quiet* as very important in the treatment, confining the child to the house for several days, and if the case is a severe one, in bed. All active measures, he advises, should be avoided, particularly inflation, and the patient cautioned against violently blowing the nose. For the immediate relief of pain, where the condition of the canal admits of deep applications being made, belladonna is found very serviceable. For this purpose, he paints the deeper parts over with a small quantity of the following:

℞. Ung. belladon. . . .
  Olei-paraffini (Vaseline) . . āā ℥ i.
M.

Or, sometimes the following is preferred:

℞. Atropiæ sulphat. . . . gr. v.
  Aquæ . . . . . ℥ i.
M. Sig. Four drops, to be dropped in the ear.

In using these, he advises that they should be first warmed, and the canal previously freed of secretions. *Dry warmth*, when grateful, is also applied with much benefit; either as heated air from hot salt bags, or by means of heated pillows. In some instances this may be all that is required. He also finds that this measure accomplishes just as much as warm water, which, though often beneficial when poured directly upon the drum head, may prove in-

jurious. Or, at times, gentle *fomentations* or steam-
ing is employed, but all active syringing, douching,
poulticing and the like, are particularly objected to;
indeed, he believes these measures often do positive
harm. In the later stages, however, syringing and
mopping is more thoroughly practiced, and occasionally the *air douche* is cautiously employed. Anodynes are seldom used by him with children; moreover, he finds that large doses of narcotics only
serve, sooner or later, to make the trouble worse.

In regard to *puncturing* the membrane when no
discharge occurs from the inflamed parts, DR. SEX-
TON advises that a resort to this procedure should
receive the most careful judgment. In his experience it is not always demanded for relief, even
where the membrana tympani is much protruded by
the accumulated secretions; as he finds other means
of treatment often efficient in causing the pain and
inflammatory symptoms to speedily subside. When,
however, there is considerable thickening and tough-
ening from previous attacks, attended with closure
of the Eustachian tube, he considers the operation
justifiable; and, when indicated, he directs that it be
done promptly, under the influence of an anæs-
thetic.

The use of *leeches* he finds unsatisfactory, as the
abstraction of blood from the adjacent part of the

cheek, or even from the concha, does not relieve the deeper congestion. Moreover, besides the irritation from the bite, their appearance tends to frighten the child, and often it is difficult to check the bleeding. *Blistering*, and painting with *iodine*, he regards as too irritating in effect, and should be avoided.

In cases of ear-ache associated with nervous excitement, he places most reliance upon pulsatilla, especially in very young children, giving it thus:

℞- Tinct. anemon. praten. . gtt. v–x.

Aquæ . . . . . ℥ iv–vi.

M. Sig. A teaspoonful as necessary.

He also uses *aconite* and *gelsemium* with great efficacy. In either case, he advises the importance of obtaining a tincture from the fresh plant.

Calcium sulphide is considered by him to be a most valuable remedy, and he rarely finds that any nausea is produced by its use. In his opinion it both prevents and arrests suppuration, or it may limit the inflammation and hasten recovery. In these cases he usually gives the drug in pill form, but where this is impracticable, it may be administered in the form of a trituration; thus:

℞. Calcii sulphidi . . . gr. ¼.

Sacchari lactis . . . gr. ii–iii.

M. Sig. Dose, every three or four hours.

Finally, in all cases of ear affections, he particu-larly directs that an examination of the mouth and upper part of the pharynx should always be made, and any causes of irritation there present receive careful treatment ; *e. g.* caries and other disorders of the teeth and gums, naso-pharyngeal catarrh, etc.

OPHTHALMIA.

———

In simple catarrhal affections of. the eye, DR. C. R. AGNEW advises that it would be much better if the general practitioner should never resort either to the solid stick, or strong solutions, of silver nitrate, or the sulphate of copper. He also cautions against the use of *poultices*, his belief, in this respect, being formulated thus : " If you want to put the eye out, poultice it." In his experience the following is the most effectual treatment in these cases :

R. Acidi tannici . . . . gr. x.
Sodii biborat. . . . gr. x.
Glycerinæ . . . . ℥ i.
Aquæ camphoræ . . . ℥ i.

M. Sig. To be applied once daily, in form of the spray.

For applying this, he uses a Davidson's atomizer, No. 55, thoroughly spraying the everted lids once daily. (After using this solution in an atomizer, it it will be necessary, each time, to wash the instrument out with water, in order to keep it in perfect working condition.) In addition to the spray, he directs that the eyes be bathed with a solution ·of salt in water, ℥ i–Oi., the temperature of the water to be determined by the feelings of the patient. Or, in many instances, he employs the following with equally good results :

℞. Sodii biborat. . . . . ℥ ii.
     Aquæ camphoræ . . . ℥ vi.
M.

Of this, a tablespoonful is to be mixed with a tablespoonful of hot water, and used for bathing the eyelids from three to five times daily. In doing this, he advises the precaution of freshly mixing the solution each time it is used ; otherwise, some patients are apt to continue with the same water over and over again.

In all cases, DR. AGNEW directs especial attention to *diet*, *hygiene*, and *exercise*. The prevalent habit of scant feeding must be corrected, and the food made of the most nutritious character, taken regularly and in gradually increasing amounts. These measures he considers of the highest practical importance.

In the purulent ophthalmia of infants, Dr. J. Lewis Smith recommends the following with much satisfaction :

℞. Hydrarg. corros. chlor. . . gr. i.

Aquæ rosæ . . . . ℥ ii.

Aquæ . . . . ℥ vi.

M. Sig. Apply every three hours.

PART VI.
SKIN DISEASES.

ECZEMA.

In infantile eczema, DR. L. DUNCAN BULKLEY advises that too much dependence must not be placed on the local treatment of the disease, as he finds that evidences of imperfect assimilation can always be discovered in these children. The evacuations from the bowels are faulty; the urine constantly presents traces of mal-assimilation, and thorough investigation invariably shows an imperfect state of health. Moreover, in nursing children, he considers it of the highest importance that careful attention be paid to the mother, who very frequently exhibits dyspepsia or constipation, or is considerably debilitated ; or possibly is taking ale, beer, unnatural quantities of tea, etc.; all of which disagree and cause trouble in the child. Therefore, if the influence of internal, general, dietary, and hygienic causes be strictly attended to, he finds that much less in the way of treatment is required locally; and what is thus used, is more rapidly and more completely successful. But he also cautions that if these are not recognized and managed with care, the results of local treatment are imperfect and uncertain.

Internal medication, to a certain degree, he considers absolutely necessary. As a rule, he gives small purgative doses of calomel every other day, and a mild alkali, such as *potassium acetate* in the liquor ammoniæ acetatis, with a little nitre, and perhaps aconite. Individual cases, however, require different management.

In regard to *mechanical restraint* to prevent the scratching, DR. BULKLEY aims to relieve the itching by proper applications, and thus avoid resorting to such extreme measures. The diachylon ointment, employed by some, he considers very inefficient for this purpose in infantile cases. Tar in some form is much preferred, the following combination being a favorite with him :

℞. Unguent. picis   .   .   .   . ℥ i.
　Zinci oxidi   .   .   .   .   . ℨ ii.
　Unguent. aquæ rosæ   .   .   . ℥ iii.
M.

This, he advises, should be carefully prepared, and very thoroughly and abundantly applied. If it appears stimulating, less of the tar ointment is used. He lays great stress on employing the *rose ointment*, and not simple cerate, or lard, or vaseline, or petroleum. He also directs that the ointment be made of a consistency to spread easily, and yet not to all melt away after application. Concerning the use of *water* to eczematous surfaces in children, he advises that they should be washed only as directed, and that very rarely; often only at intervals of several days. Furthermore, in his experience, it is all important that the protective ointment be replaced immediately after the surface is dried, and renewed sufficiently often to keep the parts completely shielded ; even twenty or more times during the first day. On covered parts, the ointment may be spread thickly on the wooly side of sheet lint, and bound on. As to the use of a *mask* for the face, he never resorts to this measure, and rarely finds it necessary to restrain the infant much after the first day or two. The only approach to this practiced by him, is putting on muslin mittens, tied about the wrist, and with tapes from these passing behind the back, or beneath one leg. Under this management, if every particular is carried out, he is assured of but one result, namely, arrest of the eruption and, if the dietetic and hygienic elements are also persisted in, a cure of the disease.

Regarding *arsenic* in the treatment of eczema, although not considered as a specific, yet he finds it of great service, on account of its effect in quieting nervous irritation. Indeed, in many instances, he has obtained almost instant relief from the internal administration of Fowler's solution alone. As a rule, in all cases, he gives arsenic internally at some stage of the disease.

In eczema of the anus and genitals, not infrequently met with in older children, and more particularly in chronic cases, DR. BULKLEY has derived great satisfaction from the following method of treatment. To correct the imperfect intestinal secretions and the imperfect liver action, which is usually an accompaniment of this condition, is, in his estimation, of primary importance. For this purpose he advises, the most careful judgment is often necessary ; purgatives and laxatives are not sufficient, nor does any routine plan answer for all. Generally, however, he finds the following prescription very useful :

R̵. Sulphur. præcipitat.   .   .   .

Potass. bitart.   .   .   .   aa ℥ i.

M. Sig. A teaspoonful at night, rubbed up with sufficient water to make a paste.

In addition, diet, hygiene, exercise, and regularity in attending to the calls of nature, are also required, together with what internal medicinal assistance may be needed. If the disease is simply due to debility, he administers *iron* and other tonics with much benefit. Arsenic he seldom, if ever, employs as a curative measure at the beginning, in these cases, and especially not in acute forms. But in the later stages, and where there is marked eczematous habit, when, after all other means have been carefully attended to, there still remains a tendency to the disease, he then uses arsenic in connection with other remedies.

Local treatment is also considered of vast importance. He particularly cautions, however, that too strong applications are apt to do more harm than good, and directs that the soothing plan be followed as far as possible, especially where there are signs of inflammation present; stimulating measures being resorted to only in the later stages. The itching, he finds, yields promptly to mild treatment, together with the proper general assistance. He places most reliance upon *hot water* (not merely warm) to relieve the congestion. To be of service, however, DR. BULKLEY insists that the following plan must be strictly adhered to. He directs that the patient be made to sit upon the edge of a chair, having a basin of water at hand with a soft handkerchief in it. This latter is then to be held in a mass, as hot as can be borne, to the parts, for a minute at a time, and the

process repeated three times. Too long or too frequent bathing, or rubbing with a cloth, is injurious.
Ordinarily, the hot water is applied only once in
twenty-four hours; usually immediately before retir-
ing thus affording a quiet and generally sound sleep.
All scratching, or even touching the part, must be
carefully avoided if possible. Before applying the
hot water, the ointment to be employed is prepared,
thickly spread on lint, and cut of a size to cover the
affected parts only. After these have been dried by
soft pressure, but absolutely without friction, the
already spread cloths are immediately applied, thus
at once and entirely excluding the air. In severe
cases the hot water is repeated occasionally, but
usually he finds it sufficient to simply renew the
ointment one or more times during the day. The
ointment employed varies somewhat with the case ;
as a rule, however, he prescribes the tar and zinc
oxide ointment with ung. aquæ rosæ, already men-
tioned. Vaseline, as a base, is objected to, as it
soaks in too rapidly ; thus leaving the parts dry and
exposed to the air Or, in other instances, the following combination is used by him very effectually:

    ℞. Unguent. picis   .   .   .   ʒ iii.
       Unguent. bellad.   .   .   .   ʒ ii.
       Tinct. aconit. rad.   .   .   .   ʒ ss.
       Zinci oxidi   .   .   .   ·   ʒ i.

Unguent. aquæ rosæ. . . ℥ iii.

M. Ft. unguent.

An ointment of chloral and camphor is also, in his experience, a very efficient anti-pruritic. This he formulates as follows:

℞. Chloral. hydrat. . . .

Camphoræ . . . āā ʒ i–ii.

Ung. aq. rosæ . . . . ℥ i.

M.

Or, in certain cases, he finds lotions sometimes very serviceable; often prescribing the following:

℞. Bismuthi subnit. . . . ʒ ii. .

Acid. hydrocyan. dil. . . . ʒ i.

Emuls. amygd. . . . ℥ iv. .

M. Ft. lotio.

This, he advises, should not be used where the ꞈkin is torn or broken. When stronger applications are necessary, as in cases where congestion has ceased, leaving some thickening and a tendency to the formation of fissures, he uses the following with good effect.

℞. Saponis viridis . . .

Olei cadini . .

Alcohol . . · . āā ℥ i.

M.

This is rubbed briskly over the parts for a few minutes, and followed immediately afterward by a mild ointment, such as:

17

℞. Zinci oxidi   .   .   .   .   ℨ ss.
    Ung. aquæ rosæ   .   .   .   ℥ i.
M.

Or :

℞. Bismuthi subnit.   .   .   .   ℨ ss.
    Ung. aq. rosæ   .   .   .   .   ℥ i.
M.   •

Or, he sometimes substitutes calomel for either
the zinc or the bismuth, with much satisfaction.   Occasionally, however, DR. BULKLEY resorts to the application of *caustic potash* in solution, followed by
soothing measures.   When the fissures still persist,
the solid stick of *silver nitrate* is, at times, employed ;
after which the part is packed with cotton.   But
these ultimate means of relief, he advises, are to be
used with caution, otherwise bad results may follow.

In the treatment of infantile eczema, DR. A.
JACOBI advises that the addition of potassium to
cow's milk is contraindicated, since it already contains too much in comparison with the breast milk.
In his experience sodium should be added, rather
than potassium.

The use of water should also be avoided.   In some
instances he finds it necessary to employ *physical
restraint*, to prevent the child from scratching.   As
a rule, also, a mask for the head and face generally
proves very serviceable.   In chronic cases, the first

indication is removal of the scab. For this purpose, he uses the following mixture very effectually:

℞. Liquor potassæ . · . ℥ i.

Olei olivæ . . . . ℥ viii--x.

M. Sig. To be applied from two to five times daily.

Or cod liver oil is sometimes substituted for the olive oil. This application soon breaks up the crusts so that they can easily be removed. In mild cases, he often finds that oil, soap, or poultices will suffice. After their removal, the surface is kept dry with soft cloth, and a new formation of scab thus prevented. An ointment is then applied, of which DR. JACOBI usually places most reliance upon the *unguentum diachylon*. For constitutional treatment, he administers *arsenic*, and iron when indicated, with marked benefit.

The following is the formula for the above ointment, as used at BELLEVUE HOSPITAL:

℞. Emplast. plumbi . . . ℥ v.

Olei olivæ . . - . . ℥ iv.

Olei lavandulæ . . . ℨ i.

M.

In infantile cases, if the child is still nursing, DR. G. H. FOX directs that the time of *feeding* be regulated so that the breast is given only at stated periods; beginning at first with intervals of one and one-half hours, and working up gradually until three hours intervene between the nursings. In older children, he insists upon strict regulation of the diet, which is usually in a poor condition, confining the patient to meat, soups, milk with and without bread, oatmeal, eggs, etc.; and stopping all tea, cof- ¹ fee, beer, spirit, candy, cakes, and the like, either at meals or between them. Plenty of exercise in the open air is also advised.

For local treatment, he first loosens the crusts by soaking them with olive oil, and after their complete removal, he then applies *ung. zinci oxidi*, by spreading it thickly upon old linen and binding it upon the part. In cases of eczema capitis, he sometimes orders the use of the following with much benefit:

    ℞. Acidi boracici . . . . ℨ ii.

        Aquæ . . . . . ℥ i.

  M.

After washing the head thoroughly with the above solution, *unguent. cadini* ℨ i–℥ i. is applied. In other instances, where the discharge is very free and abundant, matting the hair together and forming thick scabs, after washing the crusts with oil and soap, and thoroughly removing them, he generally obtains excellent results by applying the following ointment:

    ℞. Iodoformi . . . . ℨ ii.

Ung. zinci oxidi . . . ℥ iii.

M.

Or, where the eruption covers a larger surface, as the entire face and occiput, he directs that a *mask* of white flannel be made, with holes for the eyes, nose, and mouth, and with this to retain in position the cloths spread with ointment. In such cases, also, the zinc oxide ointment is often used very successfully.

For internal medication, he frequently prescribes the following with good service :

℞. Potassii acetat. . . . . gr. v.

Aquæ . . . . . . q. s.

M. Sig. Dose, three times daily, to a child of one to two years.

If the bowels are constipated, as is usually the case in infants, he administers *calomel in doses of gr. ss.* with much benefit.

In the eczema of older children, even of long standing, DR. FOX finds that as a rule, to which there are few exceptions, the disease responds readily to appropriate treatment ; and cases which have been considered as incurable, he frequently finds the easiest ones to cure. If seen in the first stage, when practicable, he directs that thin *sheet rubber* be applied to the part, or, where this cannot be procured, oil-silk. By this means the surface exudation is promoted, and the swelling and itching in a great measure relieved. His next object, as far as local treatment is concerned, is to soothe and protect the thin, tender, newly-formed epidermis (second stage), and to prevent a recurrence of the swelling, which would inevitably restore the moist condition of the eruption. This he accomplishes by the application, night and morning, of the ordinary zinc oxide oint- ment, spread on strips of cloth and secured with a muslin bandage; a rubber bandage, he advises, which proves decidedly efficacious in the early period of the treatment, would now do more harm than good. In the third stage, he applies the ointment of cade locally, and administers potassium acetate internally, with excellent effect; or sometimes unguent. dia- chyli is used with best results. Or, where the scalp is involved, he often prefers the white precipitate; thus:

℞. Hydrarg. ammoniat.   .   .   part. iv.
    Thymol   .   .   .   .   pars i.
    Oleo-paraffini (vaseline) .   .   part. xlv.
M. Ft. ung.

In chronic cases, often with the dry, erythematous form of eczema present, where the general health is impaired, DR. FOX advises that although the local means employed afford temporary relief, the erup- tion can only be permanently cured when the health of the patient has been in a marked degree restored.

He therefore resorts to the use of *tonics, alkaline diuretics*, etc., continued for some time. In these cases, also, instead of zinc oxide ointment alone, he frequently adds ten per cent. of cade. In chronic eczema of the scalp, he considers the following very desirable in many cases:

℞. Olei cadini . . . . . ℥ i.
Olei amygd. dulc. . . . ℥ iii.
M.

The effect of local treatment, however, is apt to be very uncertain.

DR. W. H. DRAPER considers *rest* very important, especially where the itching is intolerable; as the child is thus deprived of sleep at night, and the constitution becomes run down, preventing successful treatment. Therefore, by paying attention to this particular, he finds that surprisingly good results often obtain in a very short period. In these cases, as a rule, he places the child at once upon the use of *opium*, in some form, combined with the local application, on lint, of the following:

℞. Zinci oxidi . . . .
Ol. juniperi . . . . āā ʒ i.
Adipis . . . . . . ℥ i.
M. Ft. ung.

Frequently, also, a neutral salt of potassium is given as a diuretic.

DR. R. W. TAYLOR finds soothing applications in the form of powders or lotions, such as lead and opium, most serviceable in the erythematous stage. As the disease increases in age, he resorts to *stimulation* in addition. This latter measure he believes to be one of the cardinal points in the treatment of eczema ; at the same time keeping the surface well protected, while gradually continuing with the use of the stimulant application. Zinc ointment alone, he advises, will not effect a cure unless something directly curative is added ; for this purpose he finds tar very serviceable. He also believes that, in most cases, *strong solutions of potash* can be employed with advantage, and will afford relief where soaps often fail. In using them, however, he cautions that their action must be controlled so as to get the effect of strong stimulation, and at the same time prevent any inflammation.

For internal medication, in his experience *arsenic* very greatly assists the external treatment, through its power to stimulate the skin. Therefore, while he considers the golden rule to be to attend to the local measures, yet, in many cases, he finds the internal use of arsenic exceedingly beneficial.

The following is constantly used at the NEW YORK HOSPITAL :

℞. Ung. picis   .   . ℨ iv.

Zinci oxidi . . . . . ʒ i.

Cerat. simplic. . . . . ℥ iss.

M. Ft. unguent.

Also the following, as an eczema drying salve :

℞. Plumbi glycerat. . . . . ʒ i.

Ung. zinci oxidi . . . ℥ i.

M.

HERPES.

DR. G. H. FOX directs attention to the general health, etc., administering *tonics*, iron, strychnia, and quinine, according to the indications present. Many cases of a mild character, as in herpes faciæ and labialis following a common cold, where the eruption is slight, he finds require little or nothing more in the way of treatment. The *syrupi ferri iodidi* is often prescribed very effectually. As a rule, how-ever, for local measures he aims to protect the delicate walls of the vesicles from rupture, by means of some dusting powder, such as pulv. amyli, lycopodium, or, what he frequently considers better, an *ointment of zinc oxide and lycopodium*, applied liberally and then covered with a bandage to prevent rubbing of

the clothing.   If the vesicles have burst, leaving a
raw surface, he finds powders still useful, or sometimes a simple ointment is employed.   Frequently
he uses unguent. zinci oxidi, or some *anodyne lotion*
containing opium, belladonna, and camphor; or in
certain cases the following acts very beneficially:

℞.   Acidi carbolici   .   .   . gr. xii–xv.

Aquæ   .   .   .   .   ℥ i.

M.

He also prescribes the following with excellent
service;

℞.   Morphiæ   .   .   .   .   . gr. x.

Collodii flexile   .   .   .   . ℥ i.

M. Sig.   To be painted over the part.

In zoster, if the child suffers much pain, and
especially if the rest is much disturbed by it, DR.
FOX administers *opium* to relieve it.   For the neu-
ralgic pains, more particularly those remaining after
the local lesion has healed, he considers *Fowler's
solution in doses of gtt. v.* one of the best internal
remedies.   Where there is much mental worry, a
sleeping draught containing potassium bromide and
opium is generally administered.

In zoster occurring in older children, if the pains
are beyond endurance, DR. W. H. DRAPER gives
morphia to control them; combined, in some cases,
with *counter-irritation.*   Quinine is also found ser-

viceable in many instances. Or, *belladonna* locally applied, especially in combination with opium, he finds often affords relief. Where the pains continue, however, in spite of other means of treatment, the application of the *thermo cautery* along the spinal column is sometimes resorted to effectually. (This latter on the theory that the disease depends upon congestion or inflammation of the roots of the spinal nerves.)

ERYSIPELAS.

DR. A. JACOBI sometimes employs the following treatment very satisfactorily:

℞. Acid. carbol. . . . . gr. xii.

Acid. oleic. . . . . ℥ ii.

M.

This is applied to the skin which is tolerably normal around the erysipelatous patch, being thoroughly rubbed in with the finger. Only a small quantity is used at a time, and the application repeated at frequent intervals of half an hour or an hour.

DR. V. P. GIBNEY highly recommends the following, in cases of recurring naso-facial erysipelas of strumous origin, frequently met with in these children:

℞. Tinct. ferri chlor. . . . ℨ ii–iii.
Glycerinæ . . . . ℥ iss.
Aquæ . . . . · ℥ iss.

M. Sig. A teaspoonful every two hours.

At BELLEVUE HOSPITAL the use of the *compound tincture of benzoin* in the treatment of this disease, is employed with good results. The affected parts are painted with the tincture once or twice daily. As an application to the face, sheet lint is recommended, dipped in *hot lead and opium solution.* Instead of poultices, in erysipelas of the extremities, *oakum* is used soaked in hot plumb. et opii and covered well with oil-silk. By this means the moisture of the limb is retained, and a soothing influence exerted over the part.

At the PRESBYTERIAN HOSPITAL the following method of treatment has been administered very successfully, in cases of erysipelas in older children. *Tinct. ferri chloridi* is given internally in large doses and repeated every hour for six hours, in conjunction with *quinine in full doses three times daily.* Locally the application of liq. plumbi et opii is employed. In many instances this plan is attended with favorable results Where, however, there is rapid extension of the redness, *injections of carbolic acid* are often practiced, as follows:

℞. Acidi carbolici   . . .   m. xxx.

    Alcohol   .   .   .   .   ℨ ss.

    Aquæ destill.   .   .   .   ℨ i.

M.

This is injected at the upper border of the redness and repeated as indicated. After the extension is under control and improving, these are discontinued and local applications of carbolic acid resorted to. The iron and quinine are also continued. By this plan of treatment success has followed, when other means have proven altogether useless.

ERYTHEMA.

In cases of erythema multiforme, DR. L. D. BULKLEY finds the following mixture very potent in reducing the cutaneous congestion in this condition, [1] and considers it a remarkably valuable combination:

℞. Magnes. sulph.   .   .   .   .   ℨ i.

    Ferri sulph.   .   .   .   .   ℨ i.

    Acid. sulph. dil.   .   .   .   .   ℨ ii.

Tinct. gentian. . . . . ℥ i.

Aquæ . . . . . ℥ iii.

M. Sig. A teaspoonful to a tablespoonful according to age.

In erythema nodosum, DR. G. H. FOX often derives much benefit from placing the child on the internal use of *syrupi ferri iodidi gtt. v. three times
daily*, in conjunction with the local application of
unguent. zinci oxidi. When indicated, tonics are
also administered with good effect. The condition
of the stomach and intestinal canal should also receive careful attention, and any existing disorder
corrected.

URTICARIA.

DR. L. D. BULKLEY administers alkaline baths at
frequent intervals, combined with the subsequent inunction of carbolated cosmoline, as follows:

℞. Acid. carbol. . . . . gr. v.

Cosmolinæ . . . . ℥ i.

M. Ft. ung.

With these measures, for internal medication, he

also prescribes the *syrup of the hypophosphites* of soda, lime, and iron.

In chronic cases occurring in older children, where the general health is poor, DR. G. H. FOX first places the child on the following:

℞. Potassii acetat. . . . gr. x.

Aquæ . . . . . q. s.

M. Sig. Dose, three times daily, to a child of ten to twelve years.

This is continued for about a week, after which he stops this drug and gives *tinct. ferri chlor. gtt. x. three times daily.* After continuing the iron for a few days, he then orders the following:

℞. Sol. Fowlerii . . . . gtt. v.

Tinct. ferri chlor. . . . gtt. x.

Aquæ . . . . . q. s.

M. Sig. Dose, thrice daily, after meals.

In addition, a pill containing *gr.* $\frac{1}{10}$ *of aloin* is taken every morning and afternoon. He also directs the patient to remain out in the sun as much as possible, for a week or more. Under this treatment decided benefit follows. After a while, as improvement occurs, he changes the tincture of the chloride for the *ferri et potass. tart.* with good effect.

DR. A. A. SMITH recommends *sodii salicylat. gr. ii. every half-hour or hour,* with much satisfaction in the treatment of urticaria.

SCABIES.

(ITCH.)

DR. W. H. DRAPER usually employs *sulphur*, either by the bath, fumigation, or, as a rule, in the form of an ointment. By this latter method, also, he finds that any eruption of acne, which sometimes appears when the sulphur bath is used, is avoided. He considers it very important, however, that the epidermis should first be softened, in order to render the application of the ointment successful. This is accomplished by means of the hot bath, the skin being well rubbed and dried, and an ointment then applied as follows :

℞. Sulphuris . . . . ℥ iii.
    Potass. carb. . . . . ℥ i.
    Adipis . . . . . ℥ viii.
M.

The patient is afterward clothed in flannel and put to bed. It is also important, he advises, that the application of the ointment be made over the entire body, not simply to the hands, etc., otherwise the parasite migrates to other parts. The only portions of the body which can be spared are the face and neck, as these are rarely attacked. Sometimes a

single application is sufficient, but he generally considers it a safer plan to repeat the process three or
four times, when he finds this treatment usually
attended with success. In some rare instances, however, the disease fails to yield to the sulphur treatment. Where such is the case, this remedy is
stopped, and a *carbolic acid ointment* applied, thus:

    ℞. Acid. carbol. cryst. .   .   .   ℥ i.

    Cosmolinæ   .   .   .   .   ℥ xx.

    Sig. Melt each separately and mix.

This measure is often followed by a speedy sub-
sidence of the disease.

DR. G. H. FOX finds the following of most excellent service in this affection:

    ℞. Sulphur. præcip.   .   .   .

    Balsam. peruv .   .   .   .

    Potass. iodidi .   .   .   .   āā ℨ i.

    Cosmolinæ   .   .   .   .   ℨ vii.

    M. Ft. ung.

Or, in some instances, he uses the following with
good effect:

    ℞. Sulphur. loti   .   .   .   .

    Sapon. virid.   .   .   .   . āā ℥ i.

    M.

DR. L. D. BULKLEY considers sulphur almost a
specific in this affection. He directs that the child
be first washed thoroughly with yellow soap, and

18

then have the following ointment applied constantly for two or three days and nights:

R̸. Sulphuris . . . . . ℥ i.
    Styracis . . . . . ℥ ii.
    Ung. adipis . . . . ℥ i.
M.

By this means, he has frequently succeeded in curing the disease in a very short time.

The following sulphur paste is extensively employed at BELLEVUE HOSPITAL:

R̸. Sulphur. sublimat. . . . ℥ i.
    Ætheris . . . . . ℥ iii.
    Glycerinæ . . . . . ℥ i.
M.

PHTHIRIASIS.

(LOUSINESS.)

DR. G. H. FOX places the greatest reliance upon local treatment. In the milder cases of phthiriasis capitis, he finds that a small quantity of *white precipitate ointment* of full strength, rubbed, not in the scalp, but throughout the hair, will effect a cure in a few days. When a more desirable preparation is required, he uses the following:

℞. Hydrarg. corros. chlor. . . gr. iv.

Aquæ rosæ . . . . ℥ i.

M.

Or this may be made up with cologne water, or with alcohol, instead of rose water. Where, however, much laceration of the scalp is present, or any spots of eczema, the mercurial solution is, as a rule, unsafe. Another and very elegant prescription, which is highly favored by him in this affection, is the following:

℞. Hydrarg. corros. chlor. . . gr. iv.

Thymol . . . . gr. xvi.

Alcohol . . . . . ℥ i.

Ol. amygdal. amar. . . gtt. v.

M.

The oil of bitter almond added to the solution, he finds, gives it a more pleasant fragrance, which is sometimes commendable.

In chronic cases of long standing, he thinks there is scarcely any better and simpler plan of treatment, both for the destruction of the pediculi and the softening and removal of the ovi from the hairs, than the application of *kerosene oil*, freely used. In employing this remedy, he directs that the head be thoroughly rubbed with the oil, and at night covered with an oil-skin cap. In the morning the parts should be well washed with soap and water. In many of these neglected cases, he often finds severe forms of eczema capitis present, due to the irritation caused by these parasites. For this an ointment of zinc oxide is also applied, together with careful attention to diet, etc.

In regard to the necessity of cutting off the hair, when long, Dr. Fox does not consider it absolutely essential, provided pains are taken to keep it clean and well dressed.   Where, however, the hair is allowed to become matted through filth and neglect, he believes that the shears play an important part in the treatment.   This latter, he advises, applies not only to this, but also to other affections of the scalp.

TINEA FAVOSA.

(FAVUS.)

Dr. L. D. Bulkley considers *depilation* quite essential in the treatment of this disease, a view not held by all.

For this purpose the following method is a favorite with him :

℞. Ceræ flavæ	.	.	.	.	℥ iii.
Laccæ intabulis		.	.	.	℥ iv.
Resinæ	.	.	.	.	℥ vi.
Picis Burgundicæ		.	.	.	℥ x.
Gummi dammar.		.	.	.	℥ xii.

M.

This is made into sticks of various sizes. In using it, the end is heated until soft and then applied to the diseased scalp, the hair having been cut short. As soon as the wax becomes cool it is carefully withdrawn, bringing away the diseased hairs. He then uses the following wash with advantage :

| ℞. Hydrarg. bichloridi | . | . | . | gr. iv. |
| Aquæ | . | . | . | . | ℥ i. |

M.

Regarding this measure, depilation, he directs attention to the fact, that many chronic cases are due to the error of supposing the disease is in a fair way to recovery, when there are no longer any crusts to be seen. He therefore advises that it is unsafe to dismiss a patient, until every hair that has been, or is liable to be, affected has been epilated.

TINEA TONSURANS.

(RINGWORM.)

DR. G. H. FOX highly recommends the following, as of the greatest efficacy in many of these cases:

℞.   Acid. carbol.   .   .   .   gr. xlviii.
     Glycerinæ   .   .   .   .   .   ℥ ii.
     Alcohol   .   .   .   .   .
     Ol. cadini   .   .   .   .   āā ℥ iii.
     Ol. citronellæ   .   .   .   .   q. s.
M.

Chrysophanic acid is considered most efficacious by many; or, in exceptional cases, combined with carbolic acid.   In young children, however, it must be used with caution, if at all.   The following is the formula used at BELLEVUE HOSPITAL:

℞.   Acidi chrysophanici   .   .   .   gr. xx.
     Oleo-paraffini   .   .   .   ℥ iii. gr. x.
     Sig.   Melt the vaseline, and while hot add the acid, stirring till dissolved.

Or, *tincture of iodine*, painted on once or twice daily and followed by the use of white precipitate ointment, is often found to be sufficient.   Tonics are also valuable in many cases, especially *cod liver oil*, together with iron, etc. Sometimes the unguentum picis liquidum is of excellent service. This is combined at Bellevue Hospital, as follows:

℞. Picis liquidæ . . . . ℨ i.

   Potassæ . . . . . ℨ ss.

   Aquæ fervent. q. s. ad . . ℥ i.

M. Et add.

   Cerati . . . . . ℥ viii.

ALOPECIA AREATA.

(TINEA DECALVANS.)

In many instances DR. G. H. FOX finds the following plan of treatment very successful. He orders a five per cent. *oleate of mercury* to be rubbed into the scalp night and morning. Small doses of calomel are also given internally, as a laxative. After continuing this for two or three weeks, should no material benefit be derived, he substitutes *liquor ammon. fort.* for the oleate, as an application to the scalp, and, if indicated, administers cod liver oil with good effect. Or, in certain cases, where a stimulating wash is desired, the following is recommended by him:

℞. Tinct. capsici . . . . ℥ iii.

Tinct. cantharidis . . ℥ ii.

Aquæ rosæ · . . . . ℥ v.

M. Ft. lotio.

In others, where the general health is poor, he frequently prescribes:

℞. Tinct. ferri chloridi . . gtt. x.

Sol. Fowlerii . . . . gtt. v.

M. Sig. Dose, three times daily, to a child of
eight to ten years.

This is continued for a week or so, according to
indications. He then often finds the following very
satisfactory:

℞. Sapon. viridis . . . . ℨ i.

Spts. vini rect. . . . . ℨ i.

M. Sig. Use as a wash.

Also, in numerous instances, he gives small doses
of *ext. jaborandi fl.* with great advantage. As improvement takes place, this is then continued in
diminished doses for a short time afterward, with
benefit. Or, sometimes he uses pyrogallic acid with
very gratifying effect ; thus :

℞. Acidi pyrogall. . . . . ℈ iii.

Glycerinæ . . . . . ℈ i.

Aquæ . . . . . ℥ ii.

M. Ft. lotio.

PSORIASIS.

In many of these cases, DR. G. H. FOX first directs his treatment to the general health of the patient. If there is any stomach and intestinal disorder present, this is corrected by the administration of bismuth subnitrate and *rhei et sodæ.* After a better condition is attained in this direction, where the child is anæmic and with poor appetite, he then finds that a marked degree of improvement takes place under the administration of iron and cod liver oil. He gives :

℞. Syrupi ferri iodidi . . . ℨ iv.

    Olei morrhuæ . . . . ℥ iiiss.

M. Sig. A teaspoonful three times daily, to a child of six to ten years.

In conjunction with this he often derives benefit from the application of unguent. hydrarg. ammoniat. Or, where the lesions are very extensive he sometimes applies *acid. chrysophanic. pur.* covered with collodion, on certain portions, and on the rest *ol. cajeput.* This is continued as necessary, and is generally attended with success. Combined with this treatment, the internal use of arsenic is frequently prescribed by him with much service ; thus :

℞. Liq. potass. arsenit. . . . ℨ ii.

Aquæ cinnam. . . . ℥ iiss.

M. Sig. Half a teaspoonful three times daily.

Or, at times, the following is preferred:

℞. Sol. Fowlerii . . . . ℨ ii.

Tinct. cinchonæ co. . . ℥ iii.

M. Sig. A teaspoonful thrice daily, to a child of eight to ten years.

By these means of treatment the results are usually very favorable.

PITYRIASIS.

In pityriasis capitis, DR. G. H. FOX sometimes directs that the patient be given a hot bath every night, and during this procedure to use as a shampoo to the head and body equal parts of sapo viride and alcohol ; and upon leaving the bath, to rub vaseline into the affected parts. In certain cases, where an elegant application is desired, the following is highly recommended by him:

℞. Chloralis . . . . . ℨ iv.

Glycerinæ . . . . . ℨ i.

Spiritus myrciæ   .   .   .   ℥ x.

Aquæ   .   .   .   .   .   ℥ v.

M. Ft. lotio.

In pityriasis versicolor, he finds the following very effectual :

℞. Sulphuris .   .   .   .   .   ℨ ii.

Sapon. viridis   .   .   .   .   ℨ iii.

Glycerinæ   .   .   .   .   ℨ ss.

M. Ft. ung.

PURPURA HÆMORRHAGICA.

(MORBUS MACULOSUS.)

DR. A. JACOBI directs attention to this condition occurring in childhood, as generally arising from a defective development, from an abnormal state of the bloodvessels; often associated with a large fatty liver, with a luxuriant growth of the adipose tissue of the body, and not infrequently with fatty degeneration of the heart. One of the chief elements of his treatment is to prevent relapses of the hemorrhages, by careful attention to *dietetic and hygienic regulations.* He advises that the patient should have three or four hours leisure daily, combined with exercise out of doors. *e. g.* The child should be out in the open air for two or three hours, walk about three miles a day, and have a cold bath, or cold sponging, every morning. Regarding diet, he directs that all fatty and fat producing foods, such as fat itself, sugars, starches, etc., be rigidly avoided, and the patient fed with meat and milk in abundance, and with but a moderate amount of vegetable diet. As for medication, he derives much benefit from placing the child on the *syrup of the iodide of iron.* In some cases, the tendency to hemorrhage may be lessened by the administration of *ergot ;* but where there has been no tendency to dangerous hemorrhages, he does not consider this necessary. He also administers *dilute sulphuric acid* as a direct tonic to the blood vessels ; cautioning, however, against a too long continuance of its use, as after a time it tends to disintegrate the red blood corpuscles. Of this, he usually gives ℥ss. in the course of twenty-four hours, to a child of ten to fourteen years. In addition to these measures, in many instances, he also prescribes *quiniæ sulphat. in doses of gr. ii–iii. three times daily*, with excellent effect.